Universitext

T0216599

Springer

London
Berlin
Heidelberg
New York
Barcelona
Hong Kong
Milan
Paris
Santa Clara
Singapore
Tokyo

Universitext

Editors (North America): S. Axler, F.W. Gehring, and K.A. Ribet

(continued after index)

Graham Everest   Thomas Ward

Heights of Polynomials and Entropy in Algebraic Dynamics

With 23 Figures

 Springer

Graham Everest
School of Mathematics
University of East Anglia
Norwich NR4 7TJ, UK

Thomas Ward
School of Mathematics
University of East Anglia
Norwich NR4 7TJ, UK

Mathematics Subject Classification (1991): 11G07, 11S05, 22D40

ISBN 978-1-84996-854-6

British Library Cataloguing in Publication Data
Everest, Graham
   Heights of polynomials and entropy in algebraic dynamics. - (Universitext (UTX))
   1. Arithmetical algebraic geometry 2. Polynomials 3. Entropy
   4. Differentiable dynamical systems   I. Title II. Ward, Thomas   512.1

Library of Congress Cataloging-in-Publication Data
Everest, Graham, 1957-
   Heights of polynomials and entropy in algebraic dynamics / Everest
Graham and Thomas Ward.
      p.   cm. — (Universitext)
Includes bibliographical references and index.

   1. Arithmetical algebraic geometry. 2. Measure theory.
3. Differentiable dynamical systems. 4. Curves, Elliptic. I. Ward, Thomas, 1963-   . II. Title.
QA242.5.E94   1999   98-49603
516.3'5—dc21   CIP

12/3830-543210 Printed on acid-free paper

He determines the number of the stars
 and calls them each by name.
Great is our Lord and mighty in power;
 his understanding has no limit.

Psalm 147, verses 4–5.

Preface

Arithmetic geometry and algebraic dynamical systems are flourishing areas of mathematics. Both subjects have highly technical aspects, yet both offer a rich supply of down-to-earth examples. Both have much to gain from each other in techniques and, more importantly, as a means for posing (and sometimes solving) outstanding problems. It is unlikely that new graduate students will have the time or the energy to master both. This book is intended as a starting point for either topic, but is in content no more than an invitation. We hope to show that a rich common vein of ideas permeates both areas, and hope that further exploration of this commonality will result.

Central to both topics is a notion of complexity. In arithmetic geometry 'height' measures arithmetical complexity of points on varieties, while in dynamical systems 'entropy' measures the orbit complexity of maps. The connections between these two notions in explicit examples lie at the heart of the book. The fundamental objects which appear in both settings are polynomials, so we are concerned principally with heights of polynomials. By working with polynomials rather than algebraic numbers we avoid local heights and p-adic valuations.

Our approach is computational and examples are included to illustrate the basic notions. There are 103 exercises in the text (some well-known and some original). Since the background required for some of them goes a little beyond the text, hints at solutions for all of them are included. Understanding this additional material, and expanding the solution hints to complete arguments, is a 'meta-exercise'. There are also 14 'questions': these are either open-ended suggestions or known open problems. Some of the main results are proved under simplifying assumptions (for example, certain results about many-variable polynomials are proved for two variables), and some stated theorems are not proved at all. This is part of a deliberate attempt to advertise the subjects without losing the reader in a mass of detail.

WHAT THE BOOK CONTAINS. Starting with Lehmer's beautiful paper from 1933, we sketch some of the history of what has become known as Mahler's measure of a polynomial. In his paper, Lehmer constructed large primes in a way that generalized the Mersenne primes (which is also a topic of current interest), building on earlier work of Pierce and others. Here we already see a coincidence of ideas appearing: all of his basic ingredients arise as dynamical data. Lehmer's famous problem (to bound below his measure of growth in a uniform way) can be interpreted in dynamical terms. In the first chapter, we describe some partial results on Lehmer's problem. These have been selected to satisfy two criteria: each is definitive for its own special case, and each proof offers great aesthetic appeal with little technical anguish. Thus Schinzel's lower bound for the measure of integral polynomials with real zeros is included because it does both. The same applies to Smyth's resolution of Lehmer's problem in the non-reciprocal case (though we present a simpler weaker version), and to Zagier's proof of Zhang's theorem. Dobrowolski's result is not included: this is elegant and important but was excluded because it is unclear whether it is the last word on the general case. Similarly, we omit further harder results of Dobrowolski, Schinzel and Lawton bounding the non-zero measures in terms of the number of non-zero coefficients. Clearly the selection has involved painful choices and we hope an interested reader would go on into the wider literature. Some references to the extensive literature on Mahler's measure have been included, but we make no claim to completeness or to historical accuracy concerning the literature.

The second chapter introduces dynamical systems associated to integral polynomials at two levels. The first deals with toral endomorphisms (coming from monic polynomials), and some proofs are merely illustrated by examples. The second deals with automorphisms of 'solenoids' and complete proofs are given at the expense of greater technical demands. In both settings the objects from Chapter 1 appear again in dynamical terms. A central result here is that the integer sequences studied by Lehmer arise as periodic points, and the Mahler measure appears as an entropy. The notion of expansiveness for dynamical systems is introduced, and it is shown that Lehmer's growth rate measure only exists for expansive systems. This is preparatory material for Chapter 4, which is an account of certain dynamical systems arising from polynomials in several variables.

In Chapter 3, we present some truly beautiful mathematics concerned with generalizing notions in Chapter 1 to the many variable case. It is here that Mahler's insight has proved to be so valuable: he noticed (in 1960) that Lehmer's measure has an integral representation, and we have dubbed this 'Mahler's lemma'. Despite being simple (a direct application of Jensen's formula from complex analysis) it is of fundamental importance as it allows a natural definition of measure for polynomials in several variables. The chapter is devoted to three main topics. The first is to present some of the fascinating specific calculations over the years, including some recent ones. In particular we state some of Boyd's recent examples (motivated by work of Deninger) where Mahler measures for two-variable polynomials arise as values of L-functions of elliptic curves. The original intention to include some material on this connection was abandoned because of the massive background required to even state the basic conjectures. We go on to give Smyth's geometric proof of the classification of polynomials with vanishing measure. Finally, we give a thorough account of Lawton's ingenious proof of a limit formula which shows that many-variable measures are limits of single-variable measures.

Chapter 4 describes the dynamical system associated to a polynomial in several variables. The periodic points give an analogue of Lehmer's original growth rate for polynomials in several variables. The entropy formula is described, with some examples discussed in detail. We also include some more recent work by Lind on dynamical zeta functions for higher-dimensional systems.

The first of the two chapters on elliptic curves, Chapter 5, contains mainly classical material. Eleven years before Lehmer's paper, Mordell had used a notion of height to prove his famous and influential theorem on the finite generation of the group of rational points on an elliptic curve. We give a thorough account of how the notion of height is used to prove this, assuming for brevity the weak Mordell theorem. We go on to develop the more functorial notion of the canonical height, and go far enough to make it clear that the canonical height is some kind of analogue of Mahler's measure.

By this point, one theme should become clear. There is a linkage between Mahler's measure and arithmetic dynamical systems. There is another linkage between Mahler's measure and the canonical height. It is reasonable to ask if this triangle of ideas can be filled in by constructing a family of 'elliptic' dynamical systems whose topological entropy is related to the canonical height and whose ergodicity is related to the (non-)vanishing of the canonical height. Chapter 6 therefore serves two purposes. One is to give a brief and self-contained introduction to the theory of elliptic curves over the complex numbers, sufficient to fix the idea that a complex elliptic curve is an analogue of the circle. The other is to make a deliberate attempt to reconcile the notions of Mahler's measure and the canonical height. In practice, this means reinterpreting the theory of the archimedean local height as it appears in Lang and Silverman (and as worked out originally by Tate). This local height is only one component of the global height, but all the ideas we want to illustrate arise here. The full picture for the possible elliptic dynamical systems is probably adelic in character, but the difficulties all lie at the archimedean place (and those coming from primes of bad reduction). Also, at the archimedean place we have all the classical functions at our disposal. We develop elliptic analogues of Lehmer's original quantities and show they have the expected properties. In particular, we note that (for real points on real curves) it is possible to realize the values of the elliptic Mahler measure as the entropy of a transcendental map (in the rational case) whose periodic points are counted asymptotically by the division polynomial evaluated at the point.

This we hope completes the circle of ideas in a reasonable way for a book at this level.

READERSHIP AND BACKGROUND. The book can be read by postgraduates (or advanced undergraduates). It arose out of lectures given to postgraduate and MMath students at the University of East Anglia. With a few exceptions (notably the Nullstellensatz in Chapter 5, Haar measure, elementary abelian harmonic analysis and the Gelfand transform in Chapters 2 and 4) the only background required is the typical mix of algebra and complex analysis taught in most British universities. Thus basic ring and group theory are assumed, as are complex analysis as far as Cauchy's theorem, elementary Fourier analysis, residue calculations and some understanding of uniform and absolute convergence of power series. Three appendices contain less standard material, a fourth contains a simple proof of an important estimate due to Mahler, a fifth contains some calculations by Bríd ní Fhlathúin, a sixth hints for the exercises, and the last appendix contains a glossary of symbols.

RELATIONSHIP BETWEEN CHAPTERS. Chapters 1 and 3 are self-contained and comprise an introduction to the Mahler measure of polynomials. Chapters 5 and 6 are self-contained and comprise an introduction to elliptic curves, canonical heights and the elliptic Mahler measure. Chapters 2 and 4 use some material from 1 and 3 respectively, but are otherwise a fairly self-contained introduction to the simplest possible family of algebraic dynamical systems.

SOURCES. Much in this book is in the literature, and there are excellent texts on some of the topics considered. For elliptic curves and canonical heights, there are the texts of Silverman [Sil86], [Sil94], although we interpret the height in a new way. There is a useful chapter on elliptic functions from an elementary point of view in Whittaker and Watson [WW63, Chapter XX]. For algebraic dynamical systems there is Schmidt's monograph [Sch95] (which also includes some of the results on polynomials and Mahler measures), and for a general introduction to the field of dynamical systems and measures of orbit complexity there is the comprehensive introduction by Katok and Hasselblatt [KH95]. For polynomials there is Schinzel's book [Sch82]. The material on Mahler's measure (both classical and elliptic) is taken from papers, and much of it has not appeared in book form before. For background on ergodic theory and dynamical systems, there are the books of Walters [Wal82] or Petersen [Pet83].

NOTATION AND NUMBERING. Notation is explained in Appendix G. Theorems, lemmas, definitions, equations, and remarks are numbered $X.Y$ for the Yth statement in Chapter X. Exercises are numbered separately by chapter, and questions are numbered separately independent of chapter. Bold numbers in the index indicate definitions or main entries.

ACKNOWLEDGMENTS. Besides the patient support of Susan and Tania, we would like to acknowledge the help received from David Boyd, Manfred Einseidler, Wayne Lawton, Andrezj Schinzel, Joe Silverman and Christopher Smyth. Also, for the careful reading of sections of the manuscript at various stages, we thank Paola D'Ambros, Bríd ní Fhlathúin, Richard Miles, Christian Röttger and the class of MTH4A22 (Spring 1998) at UEA.

Norwich, September 1998   *Graham Everest and Thomas Ward*

Contents

1. Lehmer, Mahler and Jensen

1.1 Background: Lehmer's Paper

In 1933, D.H. Lehmer published a paper [Leh33] entitled 'Factorization of certain cyclotomic functions'. A by-product of his factorization method was a way of manufacturing large primes: 'large' should be interpreted in the light of available computing machinery in 1933 (see Section 1.6 below). The construction was to take a monic, integral polynomial

$$F(x) = x^d + a_{d-1}x^{d-1} + \ldots + a_1 x + a_0 \in \mathbb{Z}[x],$$

with factorization over $\mathbb{C}$

$$F(x) = \prod_{i=1}^{d} (x - \alpha_i).$$

Then, for every $n \in \mathbb{N}_{>0}$ define

$$\Delta_n(F) = \prod_{i=1}^{d} (\alpha_i^n - 1). \tag{1.1}$$

It is better to assume that no α_i is a root of unity (since if $\alpha_i^N = 1$ for some N, then $\Delta_n(F) = 0$ for all n divisible by N). In any case, the quantity $\Delta_n(F)$ is always an integer since the product (1.1) contains all the algebraic conjugates of all the zeros of F (see Remark A.5 in Appendix A for the details; alternatively, in Lemma 2.3 we show that $\Delta_n(F)$ is the cardinality of a finite group). Lehmer was able to produce some large primes as values of $\Delta_n(F)$. Substantial work on these sequences was also done by T.A. Pierce in his 1917 paper [Pie17], where the form of prime factors of $\Delta_n(F)$ was described.

Example 1.1. Let $F(x) = x^3 - x - 1$. Lehmer showed that

$$\Delta_{113}(F) = 63,088,004,325,217$$

and

$$\Delta_{127}(F) = 3,233,514,251,032,733$$

are primes.

The motivation in [Leh33] and [Pie17] may have been to generalize the classical notion of Mersenne number and Mersenne prime – Lehmer worked on related problems before and after this paper (see [BLS75], [BLS+83], [Leh30], [Leh32], [Leh47]). The Mersenne prime case is achieved by choosing the polynomial

$$F(x) = x - 2,$$

so that

$$\Delta_n(F) = M_n = 2^n - 1.$$

In his paper, Lehmer demonstrated that $\Delta_n(F)$ is more likely to produce primes if it does not grow too quickly, and measured the rate of growth by considering the ratio of successive terms,

$$\left| \frac{\Delta_{n+1}(F)}{\Delta_n(F)} \right|. \tag{1.2}$$

Lemma 1.2. *Provided no root α_i of F has $|\alpha_i| = 1$,*

$$\lim_{n \to \infty} \left| \frac{\Delta_{n+1}(F)}{\Delta_n(F)} \right| = \prod_{i=1}^{d} \max\{1, |\alpha_i|\}. \tag{1.3}$$

Proof. This is clear since we can treat each term in the product separately:

$$\lim_{n \to \infty} \left| \frac{\alpha^{n+1} - 1}{\alpha^n - 1} \right| = \begin{cases} |\alpha| & \text{if } |\alpha| > 1, \\ 1 & \text{if } |\alpha| < 1. \end{cases}$$

Exercise 1.1. [1] Give examples to show that in general an integral polynomial may have zeros with unit modulus which are not unit roots.
[2] Show that a monic example of [1] can only occur in degree at least 4.

Exercise 1.2. Show that $\Delta_n(F)$ is a *divisibility sequence*. That is, prove that if n divides m then $\Delta_n(F)$ divides $\Delta_m(F)$.

Remark 1.3. Lehmer made the following remark concerning the non-trivial problem of the convergence of (1.2) in the presence of possible unit modulus zeros (he assumed that F was irreducible, and defined Ω to be the right-hand side of (1.3): 'It may happen that F has a root α on the unit circle. For $|\alpha| = 1$, (1.2) contributes an oscillating factor which, although it never vanishes or becomes infinite (since F is not a cyclotomic function), cannot be estimated readily. For lack of something better we use Ω to measure the rate of increase of the sequence

$$\Delta_1, \Delta_2, \Delta_3, \ldots$$

even when some of the zeros of F lie on the unit circle.' In fact the expression (1.2) only converges if there are no unit modulus zeros, whereas the expression used in Lemma 1.10 below always converges.

1.2 Mahler's Measure

Thirty years later, Mahler used a generalization of the measure (1.3) of a polynomial in a quite different setting (see Remark 1.5 below).

Definition 1.4. For any non-zero polynomial

$$F(x) = a_d x^d + a_{d-1} x^{d-1} + \ldots + a_0 = a_d \prod_{i=1}^{d} (x - \alpha_i).$$

in $\mathbb{C}[x]$, define the *Mahler measure* of F to be

$$M(F) = |a_d| \cdot \prod_{i=1}^{d} \max\{1, |\alpha_i|\}.$$

In this definition, an empty product is assumed to be 1 so the Mahler measure of the non-zero constant polynomial $F(x) = a_0$ is $|a_0|$.
    Write

$$m(F) = \log M(F)$$

for the *logarithmic Mahler measure*, and extend the definition to include $m(0) = \infty$. This convention looks a little strange, but makes sense in the dynamical interpretation: see Exercise 2.4 and Theorem 2.12 in Chapter 2. Since we often use $m(F)$, it will also be called the Mahler measure below.

Remark 1.5. This measure is called the Mahler measure because of two papers written by Mahler in the early 1960s – [Mah60] and [Mah62]. His interest in the quantity $M(F)$ was not to study it for its own merits, but instead to compare it with other kinds of measures – more natural ones in some sense. For a polynomial $F(x) = a_d x^d + \ldots + a_1 x + a_0 \in \mathbb{C}[x]$, define

$$H(F) = \max_{0 \leq 1 \leq d}\{|a_i|\}, \qquad L(F) = \sum_{i=0}^{d} |a_i|,$$

the *height* and *length* of F respectively. Mahler proved that

$$|a_i| \leq \binom{d}{i} M(F) \text{ for all } i = 0, \ldots, d \tag{1.4}$$

and also showed that all three measures are commensurate in the sense that

$$H(F) \ll M(F) \ll H(F) \tag{1.5}$$

and

$$L(F) \ll M(F) \ll L(F), \tag{1.6}$$

with the implied constants depending only on the degree d (see Appendix G for the notation $\ll$). Mahler [Mah64] also related the measure to the discriminant of the polynomial. The absolute value of the discriminant of F is defined to be

$$|\Delta(F)| = |a_d|^{2d-2} \prod_{i \neq j} |\alpha_i - \alpha_j|$$

where $F(x) = a_d \prod_{1 \leq i \leq d}(x - \alpha_i)$. Mahler showed that

$$|\Delta(F)| \leq d^d M(F)^{2d-2}. \tag{1.7}$$

Exercise 1.3. [1] Prove that

$$-d \log 2 + \ell(F) \leq m(F) \leq \ell(F),$$

where we write $\ell = \log L$. This is equivalent to an exact description of the implied constants in (1.6) above:

$$2^{-d} L(F) \leq M(F) \leq L(F).$$

[2] Give examples to show that the inequalities in part [1] and in (1.4) cannot be improved in general.

[3] Prove a weaker form of the inequality (1.7) as follows. Assume that

$$F(x) = x^d + a_{d-1}x^{d-1} + \ldots + a_0 = \prod_{1 \leq i \leq d}(x - \alpha_i)$$

is monic, so the absolute value of the discriminant is

$$|\Delta(F)| = \prod_{i \neq j} |\alpha_i - \alpha_j|.$$

Prove that

$$|\Delta(F)| \leq 2^{d(d-1)} M(F)^{2d-2}.$$

The inequality (1.5) requires some later material and is given in Exercise 1.11 below.

Definition 1.6. A polynomial is *cyclotomic* if all the zeros are roots of unity. The word cyclotomic means literally 'circle dividing' and it refers to the way that roots of unity divide up the unit circle.

Exercise 1.4. [1] Let $F \in \mathbb{Z}[x]$ denote a monic irreducible polynomial of degree d with zeros $\alpha_1, \ldots, \alpha_d$. Prove that, for any prime p,

$$p^d \leq \left| \prod_{i,j=1}^{d} (\alpha_i^p - \alpha_j) \right| \tag{1.8}$$

provided F is not cyclotomic.

[2] Let p denote a prime with

$$eL(F) < p < 2eL(F)$$

(which exists by Bertrand's postulate). Use this prime and (1.8) to deduce that

$$M(F) > 1 + \frac{\log 2e}{2e} \cdot \frac{1}{L(F)} \tag{1.9}$$

when F is irreducible and not cyclotomic. This exercise is based on Dobrowolski [Dob81, Lemma 2], where it is the first step towards a much deeper result. It follows from (1.9) that if F varies over a sequence of non-cyclotomic irreducible polynomials in $\mathbb{Z}[x]$ with $L(F)$ uniformly bounded *above*, the resulting non-zero values of $M(F)$ are uniformly bounded *below*.

Remark 1.7. [1] The polynomial $F(x) = x^3 - x - 1$ used in Example 1.1 has turned out to be very special. Among a certain infinite family of polynomials – the *non-reciprocal* polynomials – its measure is known to be minimal. See Smyth [Smy71] for the details and the survey paper by Boyd [Boy81] for an overview. We prove a weaker version of this result in Theorem 1.19 below.

[2] The quantity $|\Delta_n(F)|$ has an important interpretation in the theory of dynamical systems (see Section 2.2.1 below).

[3] In his paper, Lehmer mentioned that he could find no smaller measure of growth than that of the polynomial

$$G(x) = x^{10} + x^9 - x^7 - x^6 - x^5 - x^4 - x^3 + x + 1, \tag{1.10}$$

and that is still the smallest known example. Of the ten zeros of (1.10), eight lie on the unit circle and just one lies outside.

The problem of verifying that integral polynomials have a smallest positive measure is now known as 'Lehmer's problem', and it seems to be a very deep problem. See Waldschmidt [Wal80], Boyd [Boy81] and Stewart [Ste78b] for surveys of this problem. In Sections 1.3, 1.4 and 1.5 below we show how versions of this problem may be solved for certain classes of polynomials. Lehmer's problem arises in many different areas. In algebraic dynamical systems it is related to the existence of algebraic models for certain abstract dynamical systems (see the discussions after Definition 2.7 and Theorem 4.2). Lehmer's problem also turns up in statistical mechanics (see Moussa [Mou83], [Mou90]; Barnsley, Bessis and Moussa [BBM79]) and in the study of iteration of complex functions (see Moussa [Mou86]; Moussa, Geronimo and Bessis [MGB84]).

The current best unconditional results on Lehmer's problem itself are probably the following. Blanksby and Montgomery [BM71] showed that if $F \in \mathbb{Z}[x]$ has $m(F) \neq 0$ and degree d, then

$$m(F) \geq \log \left(1 + \frac{1}{52d \log 6d}\right). \tag{1.11}$$

Dobrowolski, [Dob79] showed under the same hypotheses that

$$m(F) > \frac{1}{1200} \left(\frac{\log\log d}{\log d} \right)^3.$$

(1.12)

A similar estimate was also obtained independently by Cantor and Strauss [CS82], and Rausch [Rau85] improved the bound for large values of d. Louboutin [Lou83] strengthened the result, again for large d, and Voutier [Vou96] has proved a similar result for *all* values of d. In a different direction, Dobrowolski [Dob81] proved that

$$m(F) > \frac{\log 2e}{2e(k+1)^k}$$

(1.13)

if F is a non-cyclotomic irreducible polynomial in $\mathbb{Z}[x]$ with k non-zero coefficients.

Dobrowolski, Lawton and Schinzel [DLS83] proved that if $F \in \mathbb{Z}[x]$ has $m(F) \neq 0$ and has k non-zero coefficients, then

$$m(F) > C = C(H(F), k),$$

(1.14)

where $H(F)$ is the maximum of the absolute values of the coefficients of F (cf. Remark 1.5). They also proved a bound of the form

$$m(F) > C(k)$$

(1.15)

involving only the number of non-zero cofficients. In Exercise 1.4 above some simple steps in the direction of the bounds involving only the number of non-zero coefficients are given. Dobrowolski [Dob91] gave an improved bound of the same form, showing that if F is a monic polynomial with $F(0) \neq 0$ that is not a product of cyclotomic factors then

$$M(F) \geq 1 + \frac{1}{a \exp(bk^k)}$$

where k is the number of non-zero coefficients of F and a, b are constants with $a \leq 13\,911$ and $b \leq 2.27$.

A beautiful and important observation from Mahler's paper [Mah60] is the following.

Lemma 1.8. [MAHLER'S LEMMA] *For any non-zero $F \in \mathbb{C}[x]$,*

$$m(F) = \int_0^1 \log |F(e^{2\pi i \theta})| d\theta.$$

Proof. This is a simple consequence of Jensen's formula: for any $\alpha \in \mathbb{C}$,

$$\int_0^1 \log |e^{2\pi i\theta} - \alpha| d\theta = \log \max\{1, |\alpha|\}.$$

That is, the (potentially improper) Riemann integral exists and has the stated value. This may be applied to each term in the factorization of F over $\mathbb{C}$.

For completeness, we include a proof of Jensen's formula. For the more interesting case (where $|\alpha| = 1$) we give a standard complex analysis proof and a short real analysis proof due to Young [You86].

Lemma 1.9. [JENSEN'S FORMULA] *For any $\alpha \in \mathbb{C}$,*

$$\int_0^1 \log |\alpha - e^{2\pi i\theta}| d\theta = \log \max\{1, |\alpha|\}.$$

In the sequel it will be useful to write $\log^+ \lambda = \log \max\{1, \lambda\}$.

Proof. The statement is clear for $\alpha = 0$ so assume that $\alpha \neq 0$. First assume that $|\alpha| \neq 1$. Then

$$\int_0^1 \log |\alpha - e^{2\pi i\theta}| d\theta = \begin{cases} \log |\alpha| + \int_0^1 \log |1 - \alpha^{-1} e^{2\pi i\theta}| d\theta & \text{if } |\alpha| > 1; \\ \int_0^1 \log |1 - e^{-2\pi i\theta} \alpha| d\theta & \text{if } |\alpha| < 1. \end{cases}$$

The integral in the $|\alpha| < 1$ case may also be written (via the substitution $\theta \to -\theta$) as

$$\int_0^1 \log |1 - e^{2\pi i\theta} \alpha| d\theta.$$

It is therefore enough to prove that for any $\beta \in \mathbb{C}$ with $|\beta| < 1$,

$$\int_0^1 \log |1 - \beta e^{2\pi i\theta}| d\theta = 0.$$

Write $\Re(z)$, $\Im(z)$ for the real and imaginary parts of a complex number z. Notice that $\log |z| = \Re \log z$, so

$$\int_0^1 \log |1 - \beta e^{2\pi i\theta}| d\theta = \Re \int_0^1 \log\left(1 - \beta e^{2\pi i\theta}\right) d\theta$$

$$= \Re \int_0^1 \left(-\sum_{n=1}^{\infty} \frac{\beta^n}{n} e^{2\pi i\theta n}\right) d\theta$$

$$= \Re \left(-\sum_{n=1}^{\infty} \frac{\beta^n}{n} \int_0^1 e^{2\pi i\theta n} d\theta\right)$$

$$= 0,$$

where the summation being taken out of the integral is justified because the sum is absolutely convergent.

We are left with the case $|\beta| = 1$, for which we give two proofs.

COMPLEX ANALYSIS PROOF. The integral is now singular, so we define

$$\int_0^1 \log |e^{2\pi i \theta} - \beta| d\theta = \lim_{\epsilon \to 0} \frac{1}{2\pi i} \int_{\Gamma(\beta, \epsilon)} \frac{1}{z} \log |z - \beta| dz$$

(if this limit exists), where $\Gamma(\beta, \epsilon)$ is the contour indicated in Figure 1.1.

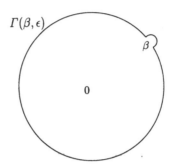

Fig. 1.1. The contour $\Gamma(\beta, \epsilon)$

Now $\frac{1}{z} \log |z - \beta| = \frac{1}{z} \Re \log(z - \beta)$, which is $\frac{1}{z}$ times the real part of a function which is analytic in the closed disc except for the point β. The residue from the singularity at 0 vanishes, so by Cauchy's theorem it does not contribute to the integral. It follows that

$$\frac{1}{2\pi i} \int_{\Gamma(\beta, \epsilon)} \frac{1}{z} \log(z - \beta) dz = \frac{1}{2\pi i} \int_{\gamma(\beta, \epsilon)} \frac{1}{z} \log(z - \beta) dz$$

where $\gamma(\beta, \epsilon)$ is the circle of radius ϵ around β. Parametrize $\gamma(\beta, \epsilon)$ by setting $z = \beta + \epsilon e^{2\pi i \theta}$ for $\theta \in [0, 1)$. Then $\frac{dz}{d\theta} = 2\pi i \epsilon e^{2\pi i \theta}$, so

$$\frac{1}{2\pi i} \int_{\gamma(\beta, \epsilon)} \frac{1}{z} \log(z - \beta) dz = \int_0^1 \frac{\epsilon e^{2\pi i \theta}}{\beta + \epsilon e^{2\pi i \theta}} \log(\epsilon e^{2\pi i \theta}) d\theta.$$

Now

$$\left| \frac{\epsilon e^{2\pi i \theta}}{\beta + \epsilon e^{2\pi i \theta}} \right|$$

is bounded, so the integral is bounded in modulus by

$$C \cdot \epsilon |\log \epsilon|$$

which goes to zero as $\epsilon \to 0$.

REAL ANALYSIS PROOF. Write the integral in the form

$$\frac{1}{2\pi} \int_0^{2\pi} \log |\alpha - e^{i\theta}| d\theta,$$

and assume that $|\alpha| = 1$; indeed, after translating by α^{-1} we may as well assume that $\alpha = 1$. Consider then

$$J = \int_0^{2\pi} \log |1 - e^{i\theta}| d\theta.$$

Since $|1 - e^{i\theta}| = 2 \sin \frac{\theta}{2}$ for $\theta \in [0, 2\pi]$, it is enough to know that

$$J = \int_0^\pi \log \sin x\, dx = -\pi \log 2.$$

This exists as an improper Riemann integral since $\sin x \sim x$ for small x (see Appendix G for the meaning of $\sim$). Write $\sin x = 2 \sin \frac{x}{2} \cos \frac{x}{2}$, then

$$J = \pi \log 2 + \int_0^\pi \log \sin \tfrac{x}{2} dx + \int_0^\pi \log \cos \tfrac{x}{2} dx.$$

Substituting $\frac{x}{2} = t$ in the first integral and $\frac{x}{2} = \frac{\pi}{2} - t$ in the second, we get

$$J = \pi \log 2 + 4 \int_0^{\pi/2} \log \sin t\, dt = \pi \log 2 + 2J.$$

This ends the proof of Lemma 1.9.

Exercise 1.5. Justify the steps in the following alternative proof of the hard case in Lemma 1.9: $\int_0^1 \log |e^{2\pi i\theta} - 1| d\theta = \int_0^1 \log |e^{2\pi i\theta} + 1| d\theta$, so

$$\int_0^1 \log |e^{2\pi i\theta} - 1| d\theta = \int_0^1 \log |e^{4\pi i\theta} - 1| d\theta$$

$$= \int_0^1 \log |e^{2\pi i\theta} - 1| d\theta + \int_0^1 \log |e^{2\pi i\theta} + 1| d\theta,$$

and therefore $\int_0^1 \log |e^{2\pi i\theta} + 1| d\theta = \int_0^1 \log |e^{2\pi i\theta} - 1| d\theta = 0$.

Another proof appears in the solution to Exercise 3.4[3].

A better (more robust) measure of the exponential growth rate of the quantity $|\Delta_n(F)|$ is

$$\lim_{n \to \infty} \frac{1}{n} \log |\Delta_n(F)|,$$

if this limit exists. The ratio (1.2) in fact *only* converges if there are no zeros with unit modulus – see Chothi, Everest and Ward [CEW97, Theorem 6.3] or Theorem 2.16 below for the details.

Lemma 1.10. *Provided no zero of F is a root of unity, the limit*

$$\lim_{n \to \infty} \frac{1}{n} \log |\Delta_n(F)|,$$

always exists for non-zero $F \in \mathbb{Z}[x]$, and the limit is $m(F)$.

Proof. First notice that

$$\log |\Delta_n(F)| = \sum_{i=1}^{d} \log |\alpha_i^n - 1|,$$

so each term can be treated separately. Recall that we write $\log^+ \lambda$ for $\log \max\{1, \lambda\}$.

If $|\alpha_i| > 1$, then

$$\frac{1}{n} \log |\alpha_i^n - 1| \longrightarrow \log |\alpha_i| = \log^+ |\alpha_i|.$$

If $|\alpha_i| < 1$, then

$$\frac{1}{n} \log |\alpha_i^n - 1| \longrightarrow 0 = \log^+ |\alpha_i|.$$

If $|\alpha_i| = 1$ then there is a subsequence $n_j \to \infty$ with the property that $\alpha_i^{n_j} \to 1$, so

$$\log |\alpha_i^{n_j} - 1| \longrightarrow -\infty.$$

The question is this: how fast does it happen? This is answered using a simple application of Baker's theorem (see Lemma 1.11 below). We claim that for any algebraic number α_i, not a unit root,

$$|\alpha_i^n - 1| > \frac{A}{n^B} \text{ for all } n, \tag{1.16}$$

for positive constants A and B, independent of n. It follows that

$$\log |\alpha_i^n - 1| = O(\log n),$$

so

$$\frac{1}{n} \log |\alpha_i^n - 1| \longrightarrow 0 = \log^+ |\alpha_i|$$

again (see Appendix G for the meaning of $O(\log n)$).

In fact the statement of Lemma 1.10 is much weaker than Baker's theorem – which is used to prove the estimate (1.16). An earlier estimate – due to Gelfond [Gel60] – is exactly equivalent to Lemma 1.10. This is pointed out in the context of the dynamical interpretation (cf. Chapter 2) in Lind's paper [Lin82, Section 4]. We have chosen to use Baker's theorem because it is more accessible in the literature and is more widely known.

We have left the proof of the estimate (1.16) to one side.

Lemma 1.11. *For $\alpha \in \bar{\mathbb{Q}}$, not a root of unity, $|\alpha| = 1$,*

$$|\alpha^n - 1| > \frac{A}{n^B}$$

for all $n \geq 1$ and constants A and B depending only on α.

Recall the notation $\ll$ (cf. Appendix G): (1.16) may be written

$$|\alpha^n - 1| \gg \frac{1}{n^B}.$$

For the proof we use Baker's theorem.

Theorem 1.12. [BAKER'S THEOREM] *Let $\alpha_1, \ldots, \alpha_r$ be algebraic numbers, $\mathbf{n} \in \mathbb{Z}^r$ an integer vector, and write $|\mathbf{n}| = \max\{|n_i|\}$. Then for any choices of the logarithm branches, if $|n_1 \log \alpha_1 + \ldots + n_r \log \alpha_r|$ is non-zero, it is bounded below by*

$$|n_1 \log \alpha_1 + \ldots + n_r \log \alpha_r| \gg \frac{1}{|\mathbf{n}|^c}, \tag{1.17}$$

where c is a constant depending only on $\alpha_1, \ldots, \alpha_r$.

Proof. Various versions of this were proved in Baker's series of papers [Bak66], [Bak67a], [Bak67b], [Bak68]. An expository treatment of the stronger inhomogeneous form of (1.17) is in [Bak75, Chapter 3].

Proof (of Lemma 1.11). We are only concerned with values of n for which α^n is very close to 1. Writing $\alpha = e^{i\theta}$ for some $\theta \in \mathbb{R}$, we see that α^n is close to 1 if and only if there is an $m \in \mathbb{Z}$ for which the real number $n\theta + 2\pi m$ is close to 0. Now

$$\alpha^n - 1 = e^{i(n\theta + 2\pi m)} - 1 \sim i(n\theta + 2\pi m) \tag{1.18}$$

for small values of $n\theta + 2\pi m$. It is sufficient therefore to find a lower bound of the right form for

$$in\theta + 2\pi im. \tag{1.19}$$

Choosing any branch of the logarithm we may write $\alpha = e^{i\theta} = e^{\log \alpha}$, and choosing a non-principal branch we may write $1 = e^{2\pi i}$. Thus (1.19) can be written in the form

$$n \log \alpha + m \log 1.$$

Since this expression cannot be zero (recall that α is not a unit root), Theorem 1.12 says that

$$|in\theta + 2\pi im| = |n \log \alpha + m \log 1| \gg \frac{1}{(\max\{|n|, |m|\})^c}.$$

On the other hand, $n\theta + 2\pi m$ is small, so n and m are close to constant multiples of each other, which shows that

$$\left| n \log \alpha + m \log 1 \right| \gg \frac{1}{|n|^c}.$$

The desired estimate follows since $|z - 1| > |\log z|$ for $|z - 1| < 1$.

Exercise 1.6. [1] Prove that

$$m(F(x)) = m(F(x^n))$$

for all $n \geq 1$.
[2] Let $F \in \mathbb{C}[x]$ have degree d. Mahler proved in [Mah61] that

$$m(F') \leq m(F) + \log d.$$

Prove that this is equivalent to the statement that for any complex numbers $\alpha_1, \ldots, \alpha_d$,

$$\int_0^1 \log \left| \sum_{j=1}^d \frac{1}{e^{2\pi i \theta} - \alpha_j} \right| d\theta \leq \log d. \tag{1.20}$$

Give examples to show that the estimate cannot be improved for all F.

A proof of (1.20) using complex analysis is given in Appendix D.

Question 1. Can you find an elementary proof of (1.20)?

Question 2. Is there a meaningful *lower* bound for $m(F')$? Any lower bound for $m(F')$ in terms of $m(F)$ must necessarily involve some dependence on the constant coefficient $F(0)$, as the example $F(x) = x - N$ shows. Experiment to find a sharp lower bound for $m(F')$. As a starting point, notice that $F(x) = x - N$ satisfies

$$m(F) + \log |d/F(0)| \leq m(F').$$

Lehmer calculated the following measures:

$$M(x^2 - x - 1) = 1.618\ldots,$$
$$M(x^3 - x - 1) = 1.324\ldots,$$
$$M(x^4 - x - 1) = 1.380\ldots,$$
$$M(x^5 - x^3 - 1) = 1.362\ldots,$$
$$M(x^6 - x - 1) = 1.370\ldots,$$
$$M(x^7 - x^3 - 1) = 1.379\ldots.$$

He also studied the 'symmetric' polynomials (we now call these polynomials *reciprocal*, see Definition 1.17 below). Among these he found

$$M(x^6 - x^4 - x^3 - x^2 + 1) = 1.401\ldots,$$

and

$$M(x^8 - x^5 - x^4 - x^3 + 1) = 1.280\ldots,$$

but found no polynomials with smaller measure than

$$M(G) = 1.176\ldots,$$

where

$$G(x) = x^{10} + x^9 - x^7 - x^6 - x^5 - x^4 - x^3 + x + 1.$$

This polynomial does indeed generate some large primes: he found that

$$\sqrt{\Delta_{379}(G)} = 37,098,890,596,487$$

is prime (another prime appears in his paper, but it would appear this is what was intended). Notice that the values of Δ_n will be squares for a symmetric polynomial, so it is natural to look for prime values of the square root.

In the next few sections partial results in the direction of Lehmer's problem are described.

1.3 Lehmer's Problem I: Schinzel's Theorem

The result in this section concerns polynomials all of whose zeros are real. This is a special case of a more general result due to Schinzel, [Sch73] concerning heights of polynomials over totally real fields.

Lemma 1.13. *For any $d \geq 1$, let $y_1, \ldots, y_d > 1$ be real numbers. Then*

$$(y_1 - 1)\ldots(y_d - 1) \leq \left((y_1 \ldots y_d)^{1/d} - 1\right)^d. \tag{1.21}$$

Proof. This is a well-known application of the convexity of $\log(\cdot)$; see Hardy, Littlewood and Polya [HLP34, Section 3.6].

Theorem 1.14. [SCHINZEL] *Suppose that $F \in \mathbb{Z}[x]$ is monic with degree d, $F(-1)F(1) \neq 0$ and $F(0) = \pm 1$. If the zeros of F are all real then*

$$M(F) \geq \left(\frac{1 + \sqrt{5}}{2}\right)^{d/2},$$

with equality if and only if F is a power of $x^2 - x - 1$.

Proof. Consider $E = \prod_{i=1}^{d} |\alpha_i^2 - 1|$; this is greater than or equal to 1 since F is monic. Now

$$E = \frac{1}{M(F)^2} \prod_{|\alpha_i|<1} |\alpha_i^{-2} - 1| \times \prod_{|\alpha_i|>1} |\alpha_i^2 - 1|.$$

By Lemma 1.13, it follows that

$$E \leq \frac{1}{M(F)^2} \left(M(F)^{4/d} - 1 \right)^d = \left(M(F)^{2/d} - M(F)^{-2/d} \right)^d.$$

Since $1 \leq E$, it follows that

$$M(F)^{2/d} - M(F)^{-2/d} \geq 1,$$

so

$$M(F) \geq \left(\frac{1 + \sqrt{5}}{2} \right)^{d/2}.$$

Exercise 1.7. Explain why the condition $F(-1)F(1) \neq 0$ must be imposed for Theorem 1.14. Where is the real zeros condition used? Prove that equality can only occur as stated in the theorem.

Corollary 1.15. *If $F \in \mathbb{Z}[x]$ has real zeros then*

$$m(F) \geq \log \left(\frac{1 + \sqrt{5}}{2} \right) = 0.481\ldots.$$

Proof. If F does not have ± 1 as leading and constant coefficient, then $m(F) \geq \log 2 = 0.693\ldots$ (cf. start of proof of Theorem 1.19), so this follows from Theorem 1.14.

Remark 1.16. For a general polynomial $F \in \mathbb{Z}[x]$ with real zeros and

$$F(0)F(-1)F(1) \neq 0$$

a similar result to Theorem 1.14 holds. Let

$$F(x) = a \prod_{i=1}^{d} (x - \alpha_i)$$

with constant coefficient $F(0) = c \neq 0$. Then $\prod_{|\alpha_i|>1} |\alpha_i| = \frac{M(F)}{a}$. Similarly $\prod_{|\alpha_i|<1} |\alpha_i| = \frac{c}{M(F)}$. Now consider

$$E = a^2 \prod_{i=1}^{d} |\alpha_i^2 - 1|,$$

which is an integer greater than or equal to 1. Rearranging the product gives

$$E = a^2 \prod_{|\alpha_i|<1} |\alpha_i|^2 \left| 1 - \frac{1}{\alpha_i^2} \right| \times \prod_{|\alpha_i|>1} |\alpha_i^2 - 1|$$

$$= \frac{a^2 c^2}{M^2} \prod_{|\alpha_i|<1} \left| 1 - \frac{1}{\alpha_i^2} \right| \times \prod_{|\alpha_i|>1} |\alpha_i^2 - 1|$$

$$\leq \frac{a^2 c^2}{M^2} \left(\left(\frac{M^2}{a^2} \cdot \frac{M^2}{c^2} \right)^{1/d} - 1 \right)^d$$

$$= \left(M^{2/d} - \frac{(ac)^{2/d}}{M^{2/d}} \right)^d$$

by Lemma 1.13. Hence, since $E \geq 1$,

$$1 \leq M^{2/d} - \frac{(ac)^{2/d}}{M^{2/d}},$$

so

$$M^{2/d} \geq 1 + \frac{(ac)^{2/d}}{M^{2/d}}.$$

Since $|ac| \geq 1$, this implies

$$1 + M^{-2/d} \leq M^{2/d}$$

and the result follows as before.

A short proof of Schinzel's theorem appears in the paper of Hoehn and Skoruppa [HS93]. Improved lower bounds appear in a paper of Flammang [Fla97]. See also Smyth's papers [Smy80] and [Smy81c] for further results in the real case.

1.4 Lehmer's Problem II: Smyth's Theorem

The polynomials that are not 'symmetric' turn out to have a uniform lower bound for their Mahler measures, so candidates for smaller Mahler measures than Lehmer's example must be among the symmetric polynomials. Boyd has carried out extensive calculations of measures for reciprocal polynomials in [Boy80], [Boy89]. It is a remarkable fact that it is sufficient to look only at polynomials of height 1 (that is, with coefficients in $\{0, +1, -1\}$; for an explanation of why this is so see Mossinghoff [Mos98, Section 3.2]). Further calculations have been done by Mossinghoff [Mos95], [Mos98]; he has also found a new limit point near 1.309 in the set of (exponential) Mahler measures of integer polynomials. The paper [MPV98] by Mossinghoff, Pinner and Vaaler explores the polynomials obtained by adding a monomial to a product of cyclotomics, giving some small examples of Mahler measures. As they point out, Lehmer's best example (1.10) is given by such a procedure:

$$G(x) = (x - 1)^2 (x + 1)^2 (x^2 + x + 1)^2 (x^2 - x + 1) - x^5.$$

Definition 1.17. Suppose $F \in \mathbb{C}[x]$ has degree d; write $F^*(x) = x^d F(x^{-1})$. Then F is *reciprocal* if $F = F^*$, and is *non-reciprocal* otherwise.

For example, Lehmer's best (smallest measure) example is reciprocal, while the polynomial $x^3 - x - 1$ is not.

In 1971 C.J. Smyth published the following remarkable theorem (see [Smy71]).

Theorem 1.18. [SMYTH] *If $F(x) \in \mathbb{Z}[x]$ is a non-reciprocal polynomial, and $F(0)F(1) \neq 0$, then*

$$m(F) \geq m(x^3 - x - 1) = \log(1.324\ldots) = 0.281\ldots$$

The condition that $F(0) \neq 0$ simply means F is not divisible by x, and $F(1) \neq 0$ means F is not divisible by $x - 1$. Clearly some condition about divisibility by $x-1$ is required for if we multiply any reciprocal polynomial by $x-1$ the measure does not change but the polynomial becomes non-reciprocal.

In his thesis, Smyth also proved a stronger result:

$$M(F) > M(x^3 - x - 1) + 10^{-4}$$

unless F is reciprocal or is the minimum polynomial of $\theta_0^{1/k}$ for some $k \geq 1$, where $\theta_0 = 1.324\ldots$ is the real zero of $x^3 - x - 1$.

We shall prove a weaker result, which is a good account of Smyth's basic method and gives a uniform lower bound for the measure of non-reciprocal polynomials. This result was known to Smyth before he proved Theorem 1.18 and has been proved independently by several people, including Stewart [Ste78a].

Theorem 1.19. *If $F \in \mathbb{Z}[x]$ is non-reciprocal and $F(0)F(1) \neq 0$, then*

$$m(F) \geq \tfrac{1}{2} \log \tfrac{5}{4} = 0.111\ldots.$$

In the proof we will need the following.

Exercise 1.8. Suppose $F \in \mathbb{Z}[x]$ is monic and irreducible, $F(1) \neq 0$, and F has a zero θ with $|\theta| = 1$. Show that F must be reciprocal. (Hint: $\bar{\theta}$ is also a root).

Proof (of Theorem 1.19). We can assume that F is irreducible, monic (since if $F(x) = a \prod(x - \alpha_i)$ then $m(F) \geq \log|a|$), $F(0)F(1) \neq 0$, and $F(0) = \pm 1$: if $F(0) \neq \pm 1$, then

$$2 \leq |F(0)| = \prod |\alpha_i| \leq M(F),$$

so $m(F) \geq \log 2 > \tfrac{1}{2} \log \tfrac{5}{4}$ (here and below write $\prod$ to denote the product taken over all the roots of F).

Notice that

$$G(z) = \frac{F(0)F(z)}{F^*(z)} = \frac{F(0)\prod(z - \alpha_j)}{\prod(1 - z\alpha_j)}$$
$$= \frac{F(0)\prod(z - \alpha_j)}{\prod(1 - z\bar{\alpha}_j)}$$

since $\{\alpha_j\} = \{\bar{\alpha}_j\}$.

It is clear that $F(1) = F^*(1)$, so if $F^* = -F$ then $F^*(1) = -F(1)$ hence $F(1) = 0$, which is impossible. We deduce that F is not identically equal to $-F^*$.

We now claim that

$$G(z) = 1 + a_k z^k + \ldots \in \mathbb{Z}[[z]],$$

convergent in some neighbourhood of zero (that $G(0) = 1$ is clear). This remark (due to Raphael Salem [Sal45]) is seen as follows. Since F is monic,

$$F^*(z) = 1 + \ldots \pm z^d,$$

so

$$\frac{1}{F^*(z)} = 1 + \ldots \in \mathbb{Z}[[z]]$$

by the binomial theorem.

Now by Exercise 1.8 we may write

$$G(z) = \frac{F(0)\prod_{|\alpha_j|<1}\left(\frac{z-\alpha_j}{1-\bar{\alpha}_j z}\right)}{\prod_{|\alpha_j|>1}\left(\frac{1-\bar{\alpha}_j z}{z-\alpha_j}\right)} = \frac{f(z)}{g(z)}$$

where f and g are holomorphic functions in an open region containing the closed unit disc.

Notice that $|z| = 1$ if and only if $\bar{z} = \frac{1}{z}$. Assume that $|z| = 1$ and consider a typical factor $\left(\frac{z-\alpha_j}{1-\bar{\alpha}_j z}\right)$ of $f(z)$:

$$\left(\frac{z - \alpha_j}{1 - \bar{\alpha}_j z}\right)\overline{\left(\frac{z - \alpha_j}{1 - \bar{\alpha}_j z}\right)} = \left(\frac{z - \alpha_j}{1 - \bar{\alpha}_j z}\right)\left(\frac{\bar{z} - \bar{\alpha}_j}{1 - \alpha_j \bar{z}}\right)$$
$$= \left(\frac{z - \alpha_j}{1 - \bar{\alpha}_j z}\right)\left(\frac{1 - \bar{\alpha}_j z}{z - \alpha_j}\right) = 1$$

so

$$\left|\frac{z - \alpha_j}{1 - \bar{\alpha}_j z}\right| = 1$$

for all j, and therefore $|f(z)| = 1$.

A similar argument applies to g. We deduce that

$$|f(z)| = |g(z)| = 1 \text{ for } |z| = 1. \tag{1.22}$$

Now write

$$f(z) = b + b_1 z + \dots$$

and

$$g(z) = c + c_1 z + \dots$$

absolutely convergent in the closed unit disc. It follows that

$$G(z) = 1 + a_k z^k + \dots = \frac{b + b_1 z + \dots}{c + c_1 z + \dots} \tag{1.23}$$

in some (smaller) disc. Now b and c are real and (without loss of generality) positive; moreover

$$b = |f(0)| = \prod_{|\alpha_j| < 1} |\alpha_j| = \frac{1}{M(F)}$$

(since $\prod |\alpha_j| = 1$) and similarly $g(0) = \frac{1}{M(F)}$ so $b = c > 0$. Now compare terms in (1.23) to see that

$$b = c = \frac{1}{M(F)},$$
$$b_1 = c_1,$$
$$\vdots$$
$$b_{k-1} = c_{k-1}$$
$$ca_k + c_k = b_k$$

If $\max\{|c_k|, |b_k|\} < \frac{b}{2} = \frac{c}{2}$, then

$$c \le |ca_k| = |b_k - c_k| \le |b_k| + |c_k| < b,$$

which contradicts $b = c$. It follows that

$$\max\{|c_k|, |b_k|\} \ge \frac{b}{2} = \frac{c}{2}. \tag{1.24}$$

Assume without loss of generality that (1.24) holds with

$$|b_k| \ge \frac{b}{2}$$

(if not, repeat the argument that follows with g replacing f, noting that $|c_k| \ge \frac{c}{2}$). Consider

$$f(z) = b + b_1(z) + \ldots + b_k z^k + \ldots$$

(where k is the power appearing in $G(z)$ in equation (1.23)). We know by (1.22) that $|f(z)| = 1$ on $|z| = 1$, so

$$\int_0^1 |f(e^{2\pi i\theta})|^2 d\theta = 1. \tag{1.25}$$

Lemma 1.20. [PARSEVAL'S FORMULA] *Suppose that $\phi : \mathbb{C} \to \mathbb{C}$ is holomorphic in an open region containing the closed unit disc, with Taylor expansion*

$$\phi(z) = e_0 + e_1 z + \ldots,$$

$e_i \in \mathbb{C}$. *Then*

$$\int_0^1 |\phi(e^{2\pi i\theta})|^2 d\theta = \sum_{i=0}^{\infty} |e_i|^2.$$

Applying Parseval's formula to f we see that

$$b^2 + |b_1|^2 + \ldots + |b_k|^2 + \ldots = 1,$$

so

$$b^2 + |b_k|^2 \leq 1.$$

On the other hand

$$|b_k| \geq \frac{b}{2},$$

so

$$\tfrac{5}{4}b^2 \leq 1.$$

Since $b = \frac{1}{M(F)}$, we deduce that

$$M(F)^2 \geq \tfrac{5}{4},$$

proving Theorem 1.19.

Remark 1.21. Smyth uses the third coefficient of G, together with a more sophisticated use of Parseval's formula to arrive at his definitive result, Theorem 1.18.

It remains to prove Parseval's formula. This is a simple application of the standard orthogonality relations for the family of functions $\{e^{2\pi i n t}\}$.

Proof (of Lemma 1.20). Notice that

$$|\phi(e^{2\pi i\theta})|^2 = \phi(e^{2\pi i\theta}) \cdot \overline{\phi(e^{2\pi i\theta})},$$

so

$$\int_0^1 |\phi(e^{2\pi i\theta})|^2 d\theta = \int_0^1 \left(\sum_{m=0}^{\infty} e_m e^{2\pi i m\theta} \right) \left(\sum_{n=0}^{\infty} \bar{e}_n e^{-2\pi i n\theta} \right) d\theta$$

$$= \int_0^1 \left(\sum_{m=0}^{\infty} \sum_{n=0}^{\infty} e_m \bar{e}_n e^{2\pi i(m-n)\theta} \right) d\theta$$

$$= \sum_{m=0}^{\infty} \sum_{n=0}^{\infty} \left(\int_0^1 e_m \bar{e}_n e^{2\pi i(m-n)\theta} d\theta \right)$$

by the absolute convergence of the Taylor series on $|z| = 1$. On the other hand

$$\int_0^1 e^{2\pi i(m-n)\theta} d\theta = \begin{cases} 1 & \text{if } m = n; \\ 0 & \text{if not.} \end{cases}$$

So the integral reduces to

$$\sum_{m=0}^{\infty} \sum_{n=0}^{\infty} \left(\int_0^1 e_m \bar{e}_n e^{2\pi i(m-n)\theta} d\theta \right) = \sum_{m=0}^{\infty} |e_m|^2$$

as required.

To close this section we give another application of Parseval's formula by proving Gonçalves' formula.

Theorem 1.22. [GONÇALVES' FORMULA] *Let $F \in \mathbb{R}[z]$ be a monic polynomial with $|F(0)| \geq 1$. Then*

$$M(F)^2 + M(F)^{-2} \leq \sum_{j=0}^{d} a_j^2, \qquad (1.26)$$

where $F(z) = z^d + a_{d-1}z^{d-1} + \ldots + a_0$.

Proof. Write

$$F(z) = \prod_{j=1}^{d} (z - \alpha_j).$$

By Parseval's formula (Lemma 1.20)

$$\sum_{j=0}^{d} a_j^2 = \int_0^1 |F(e^{2\pi i\theta})| d\theta$$

$$= \int_0^1 \left(\prod_{|\alpha_j|>1} |e^{2\pi i\theta} - \alpha_j|^2 \cdot \prod_{|\alpha_j|\leq 1} |e^{2\pi i\theta} - \alpha_j|^2 \right) d\theta$$

$$= M(F)^2 \int_0^1 \left(\prod_{|\alpha_j|>1} |e^{2\pi i\theta} - \alpha_j^{-1}|^2 \cdot \prod_{|\alpha_j|\leq 1} |e^{2\pi i\theta} - \alpha_j|^2 \right) d\theta$$

$$= M(F)^2 \int_0^1 |G(e^{2\pi i\theta})|^2 d\theta,$$

where

$$G(z) = \prod_{|\alpha_j| > 1} (z - \alpha_j^{-1}) \cdot \prod_{|\alpha_j| \leq 1} (z - \alpha_j) = z^d + \ldots \pm \left(a_0 / M(F)^2 \right).$$

Apply Parseval's formula to G to get

$$\sum_{j=0}^{d} a_j^2 \geq M(F)^2 \left(1 + \frac{a_0^2}{M(F)^4} \right) \geq M(F)^2 + M(F)^{-2}.$$

Remark 1.23. [1] The original proof is due to Gonçalves [Gon50]. The proof presented here is taken from Smyth's doctoral thesis.
[2] Schinzel [Sch82] gives a different proof starting from the observation that $F(z)F(z^{-1})$ has constant coefficient $\sum_{j=1}^{d} a_j^2$.

Exercise 1.9. Prove the complex version of Gonçalves' formula: if $F \in \mathbb{C}[z]$ has $|a_0| \geq 1$, prove that

$$M(F)^2 + M(F)^{-2} \leq \sum_{j=0}^{d} |a_j|^2.$$

The next two exercises show how lower bounds for Mahler's measure may be used to deduce irreducibility results for certain trinomials. Let $H(x)$ denote the polynomial $x^m \pm x^n \pm 1$.

Exercise 1.10. [1] Prove that H has at most one non-reciprocal factor over $\mathbb{Q}$. (Hint: use Theorems 1.18 and 1.22).
[2] Show that the reciprocal factors of H are cyclotomic.

Exercise 1.11. Prove the inequality (1.5) by finding the best possible values for the implied constants.

1.5 Lehmer's Problem III: Zhang's Theorem

In the previous section, the involution $F \mapsto F^*$ on the ring of integral polynomials was used to make non-trivial estimates for the measure of polynomials not fixed by the involution (Definition 1.17 and Theorem 1.18). In a similar spirit, we introduce another involution: for $F \in \mathbb{Z}[x]$ define

$$F_*(x) = F(1 - x). \tag{1.27}$$

It is clear that $F \mapsto F_*$ is an involution on the set of polynomials.

Theorem 1.24. [ZHANG, ZAGIER] *Let ω denote a primitive 6th root of unity. Suppose $F \in \mathbb{Z}[x]$ has degree d, and $F(0)F(1)F(\omega) \neq 0$. Then*

$$m(F) + m(F_*) \geq \frac{d}{2} \log \left(\frac{1 + \sqrt{5}}{2} \right) \tag{1.28}$$

with equality if and only if F or F_* is a power of $x^4 - x^3 + x^2 - x + 1$.

Exercise 1.12. Explain why the condition $F(0)F(1)F(\omega) \neq 0$ must be imposed for Theorem 1.24.

   This theorem was proved originally as an application of a theorem of Zhang [Zha92] in the context of heights of algebraic numbers (cf. Section 5.8). His proof used Arakelov theory and did not give the optimal lower bound in (1.28). Zagier [Zag93] gave a beautiful elementary proof, also in the context of heights, yielding the optimal bound. We follow Zagier's proof, although phrased in terms of Mahler's measure. The proof is similar in spirit to the proof of Theorem 1.19.

   Let $\omega, \bar{\omega}$ denote the roots of $x^2 - x + 1 = 0$. Theorem 1.24 will follow directly from Lemma 1.27 below. To motivate this, we first consider the special case where F and F_* are monic and both have constant term ± 1. An example is $F(x) = x^2 - x - 1$.

Lemma 1.25. *There is a constant $A \geq 1$ such that for every complex number $z \notin \{0, 1, \omega, \bar{\omega}\}$,*

$$\log |z^2 - z + 1| + 1 \leq A \left(|\log |z|| + |\log |1 - z|| \right).$$

Corollary 1.26. *Suppose F and F_* in $\mathbb{Z}[x]$ are both monic with constant term ± 1 and $F(0)F(1)F(\omega) \neq 0$. Then*

$$m(F) + m(F_*) \geq \frac{d}{2A}.$$

Proof. Apply Lemma 1.25 to each $z = \alpha_i$ to obtain

$$\sum_{i=1}^{d} \log |\alpha_i^2 - \alpha_i + 1| + d \leq A \sum_{i=1}^{d} |\log |\alpha_i|| + A \sum_{i=1}^{d} |\log |1 - \alpha_i||.$$

Since F and F_* are monic, $m(F) = \sum \log^+ |\alpha_i|$ and $m(F_*) = \sum \log^+ |1 - \alpha_i|$. On the other hand, the assumption on the constant terms means that

$$|\prod_{i=1}^{d} \alpha_i| = |\prod_{i=1}^{d} (1 - \alpha_i)| = 1$$

so

$$m(F) = \tfrac{1}{2} \sum_{i=1}^{d} |\log |\alpha_i||$$

and

$$m(F_*) = \tfrac{1}{2} \sum_{i=1}^{d} |\log |1 - \alpha_i||.$$

Thus

$$d \le \sum_{i=1}^{d} \log |\alpha_i^2 - \alpha_i + 1| + d \le A \sum_{i=1}^{d} |\log |\alpha_i|| + A \sum_{i=1}^{d} |\log |1 - \alpha_i||$$

$$\le 2A \left(m(F) + m(F_*) \right)$$

giving the required inequality.

Proof (of Lemma 1.25). Consider the function

$$f(z) = \frac{\log |z^2 - z + 1| + 1}{|\log |z|| + |\log |1 - z||}$$

for $z \in \mathbb{C} \backslash \{0, 1, \omega, \bar{\omega}\}$. As $|z| \to \infty$, $f(z) \to 1$. For values of z near the points $z = \omega$ or $\bar{\omega}$, $f(z)$ is large and negative. Finally, the function is continuous everywhere except at the intersection of the circles $|z| = |1 - z| = 1$, where it is large and negative. It follows that the function is bounded above uniformly on all of $\mathbb{C}$.

Theorem 1.24 follows from a refined version of Lemma 1.25 that makes the constant explicit and uses $\log^+ |\cdot|$ instead of $|\log |\cdot||$.

Lemma 1.27. *For any* $z \in \mathbb{C} \backslash \{0, 1, \omega, \bar{\omega}\}$,

$$\log^+ |z| + \log^+ |1 - z| \ge \frac{\sqrt{5} - 1}{2\sqrt{5}} \log |z^2 - z|$$

$$+ \frac{1}{2\sqrt{5}} \log |z^2 - z + 1| + \frac{1}{2} \log \left(\frac{1 + \sqrt{5}}{2} \right),$$

with equality holding if and only if z *or* $1 - z$ *equals* $e^{\pm \pi i / 5}$ *or* $e^{\pm 3\pi i / 5}$.

Proof (of Theorem 1.24). First notice that $e^{\pm \pi i / 5}$, $e^{\pm 3\pi i / 5}$ are the roots of $x^4 - x^3 + x^2 - x + 1 = 0$.

If F is monic, then applying Lemma 1.27 to each zero α_i in turn gives

$$m(F) + m(F_*) \ge \frac{\sqrt{5} - 1}{2\sqrt{5}} \sum_{i=1}^{d} \log |\alpha_i^2 - \alpha_i|$$

$$+ \frac{1}{2\sqrt{5}} \sum_{i=1}^{d} \log |\alpha_i^2 - \alpha_i + 1| + \frac{d}{2} \log \left(\frac{1 + \sqrt{5}}{2} \right)$$

$$\ge \frac{d}{2} \log \left(\frac{1 + \sqrt{5}}{2} \right)$$

since

$$|\prod_{i=1}^{d} (\alpha_i^2 - \alpha_i)| \text{ and } |\prod_{i=1}^{d} (\alpha_i^2 - \alpha_i + 1)|$$

are positive integers.

If F has leading coefficient a then add $2 \log |a|$ to the left-hand side of Lemma 1.24, which then becomes $m(F) + m(F_*)$. Add $\log |a|$ to the right-hand side to obtain, after writing $\log |a| = \frac{\sqrt{5}-1}{2\sqrt{5}} \log |a|^2 + \frac{1}{2\sqrt{5}} \log |a|^2$,

$$\frac{\sqrt{5}-1}{2\sqrt{5}} \log |a^2 \prod_{i=1}^{d}(\alpha_i^2 - \alpha_i)| + \frac{1}{2\sqrt{5}} \log |a^2 \prod_{i=1}^{d}(\alpha_i^2 - \alpha_i + 1)|$$

$$+ \frac{d}{2} \log \left(\frac{1+\sqrt{5}}{2} \right) \geq \frac{d}{2} \log \left(\frac{1+\sqrt{5}}{2} \right)$$

since $|a^2 \prod_{i=1}^{d}(\alpha_i^2 - \alpha_i)|$ and $|a^2 \prod_{i=1}^{d}(\alpha_i^2 - \alpha_i + 1)|$ are positive integers.

Proof (of Lemma 1.27). Define a function g by

$$g(z) = \frac{\sqrt{5}-1}{2\sqrt{5}} \log |z^2 - z| + \frac{1}{2\sqrt{5}} \log |z^2 - z + 1| + \frac{1}{2} \log \left(\frac{1+\sqrt{5}}{2} \right)$$

$$- \log^+ |z| - \log^+ |1 - z|.$$

If $|z|$ is large then $g(z)$ behaves like $-\log |z|$ and, in particular, $g(z) \to -\infty$ as $|z| \to \infty$. Similarly, if z is close to one of the points $0, 1, \omega, \bar{\omega}$ then $g(z)$ is large and negative. Away from these points, g is continuous, and so attains its maximum on some finite point or points. Off the circles $|z| = 1$ and $|1 - z| = 1$ the function is the real part of a holomorphic function, so by the maximum principle for harmonic functions (see [CKP83, p. 46] for example) the maxima must be attained on these circles (cf. Appendix D). The involutions $z \mapsto 1 - z$ and $z \mapsto \bar{z}$ preserve g, so it is enough to restrict attention to $z = e^{i\theta}$ for $0 \leq \theta \leq \pi$.

First suppose that $0 \leq \theta \leq \frac{\pi}{3}$, so $|1 - z| \leq 1$. Then

$$g(z) = \frac{\sqrt{5}-1}{2\sqrt{5}} \log \left(2 \sin \frac{\theta}{2} \right) + \frac{1}{2\sqrt{5}} \log \left(2 \cos \theta - 1 \right) + \frac{1}{2} \log \left(\frac{1+\sqrt{5}}{2} \right).$$

Write $S = 4 \sin^2 \frac{\theta}{2}$ (so that $0 \leq S \leq 1$ for $0 \leq \theta \leq \frac{\pi}{3}$). Then

$$g(z) = \frac{\sqrt{5}-1}{4\sqrt{5}} \log S + \frac{1}{2\sqrt{5}} \log(1 - S) + \frac{1}{2} \log \left(\frac{1+\sqrt{5}}{2} \right).$$

Differentiating with respect to S shows that the unique maximum of g for $S \in (0, 1)$ is attained at $S = \frac{3-\sqrt{5}}{2}$, where $g = 0$ and $\theta = \frac{\pi}{5}$.

A similar argument holds for $\frac{\pi}{3} \leq \theta \leq \pi$; here $1 \leq S \leq 4$ and

$$g(z) = \frac{-\sqrt{5}-1}{2\sqrt{5}} \log\left(2\sin\frac{\theta}{2}\right) + \frac{1}{2\sqrt{5}}\log(1-2\cos\theta) + \frac{1}{2}\log\left(\frac{1+\sqrt{5}}{2}\right)$$
$$= \frac{-\sqrt{5}-1}{4\sqrt{5}}\log S + \frac{1}{2\sqrt{5}}\log(S-1) + \frac{1}{2}\log\left(\frac{1+\sqrt{5}}{2}\right).$$

The unique maximum of g is attained at $S = \frac{3+\sqrt{5}}{2}$, where $g = 0$ and $\theta = \frac{3\pi}{5}$.

To close this section, we mention some related results. Rhin and Smyth [RS97] showed that if $H \in \mathbb{Z}[x]$ is divisible by x but is not $\pm x^n$ for any n, and $G \in \mathbb{Z}[x]$ is irreducible, then

$$m(G(H(x))) > Cd$$

for some constant C, where d is the degree of the composition $G(H(x))$.

Exercise 1.13. Prove that for any polynomial $F \in \mathbb{Z}[x]$, the polynomial FF_* can be written in the form

$$F(x)F_*(x) = G(x(1-x))$$

for some polynomial $G \in \mathbb{Z}[x]$.

Dresden [Dre98] has extended Theorem 1.24 in a different direction. If α is an algebraic integer, and F_1 is the minimum polynomial of α, F_2 the minimum polynomial of $\frac{1}{1-\alpha}$ and F_3 the minimum polynomial of $1 - \frac{1}{1-\alpha}$, then he shows that the two smallest values of

$$\frac{1}{d}\left(m(F_1) + m(F_2) + m(F_3)\right)$$

are 0 and $0.4218\ldots$, where d is the degree of α. As he points out (p. 819, *ibid.*) this has the following consequence: if $F \in \mathbb{Z}[x]$ is a polynomial of degree d with the property that the cyclic group of order three generated by the map $z \mapsto 1 - \frac{1}{z}$ is a subgroup of its Galois group, then

$$m(F) \geq \frac{d}{3}(0.4218\ldots).$$

1.6 Large Primes in 1933

Lehmer's calculations, including the 16-digit prime $3\,233\,514\,251\,032\,733$ were not aimed at generating record-breaking primes, but rather at understanding primes appearing in a novel fashion. Appendix E contains some extensions of Lehmer's calculations (and their elliptic analogues).

Essentially all large primes arise from the sequences $2^n \pm 1$, for which there are special primality tests. The famous Mersenne problem asks if $M_n = 2^n - 1$ is prime for infinitely many values of n. It is well-known that $2^n + 1$ can only be prime when n is a power of 2 and very few instances of $2^{2^k} + 1$ being prime are known; however $2^n + 1$ is sometimes a product of a small factor and a large prime. For comparison, Table 1.1 shows some record primes, breaking off with the largest prime to be found without the use of electronic computing machines and ending with the current largest known prime, found by Clarkson, Woltman and Kurowski as part of GIMPS. Some of the information in this section is taken with permission from Chris Caldwell's Prime Page on the world wide web at http://www.utm.edu/research/primes/.

Notice that Robinson's – and all subsequent – calculations were performed on a computer. The computer age, far from killing the subject off, seems to have caused a revival. Laura Nickel and Curt Noll were at high-school when they discovered their record-breaking prime. The Euler–Fermat theorem, which was generalized by Pierce [Pie17] and Lehmer [Leh33], is described below. Table 1.1 is far from complete – see Ribenboim [Rib95b] for more details. Large primes of other forms (notably $(2^n + 1)/3$) continue to be studied using methods from Elliptic Curve theory (see Bateman *et al.* [BSW89] and Morain [Mor90]).

Number	Digits	Year	Prover and Method
$2^{17} - 1$	6	1588	Cataldi; trial division
$2^{19} - 1$	6	1588	Cataldi; trial division
$2^{31} - 1$	10	1722	Euler; Euler–Fermat theorem
$(2^{59} - 1)/179\,951$	13	1867	Landry; Euler–Fermat theorem
$2^{127} - 1$	39	1876	Lucas; Lucas–Lehmer test
$(2^{148} + 1)/17$	44	1951	Ferrier; Proth's theorem
$\vdots$	$\vdots$	$\vdots$	$\vdots$
M_{2281}	687	1952	Robinson; Lucas–Lehmer test
M_{11213}	3376	1963	Gillies; Lucas–Lehmer test
M_{21701}	6533	1978	Nickel & Noll; Lucas–Lehmer test
M_{86243}	25 962	1982	Slowinski; Lucas–Lehmer etc.
M_{216091}	65 050	1985	Slowinski; Lucas–Lehmer etc.
M_{859433}	258 716	1994	Slowinski & Gage; Lucas–Lehmer etc.
$M_{3021377}$	909 526	1998	Clarkson, Woltman & Kurowski; GIMPS

Table 1.1. A brief history of large primes.

Lemma 1.28. [EULER–FERMAT THEOREM] *If p and q are odd primes, and p divides $2^q - 1$, then $p \equiv 1 \pmod{q}$ and $p \equiv \pm 1 \pmod 8$.*

Exercise 1.14. Prove the Euler–Fermat theorem.

The Lucas–Lehmer test is the following.

Theorem 1.29. [LUCAS–LEHMER TEST] *Let p be an odd prime. Then the Mersenne number $2^p - 1$ is prime if and only if $2^p - 1$ divides S_{p-1}, where $S_{n+1} = S_n^2 - 2$ and $S_1 = 4$.*

Exercise 1.15. Prove Theorem 1.29.

The last result mentioned above is part of a long list of results for numbers of special forms.

Theorem 1.30. [PROTH'S THEOREM, 1878] *Let $n = h \cdot 2^k + 1$ with $2^k > h$. If there is an integer a such that $a^{(n-1)/2}$ is congruent to -1 (mod n), then n is prime.*

This is a special case of a more general result; see Ribenboim [Rib95b] for the whole story.

GIMPS (the Great Internet Mersenne Prime Search), founded by George Woltman and others, is a very efficient system for using idle time on many different computers scattered all over the world to perform a coordinated search for Mersenne primes.

Question 3. Let $F(x) = x^3 - x - 1$. Are there infinitely many primes in the sequence $\Delta_n(F)$? For some calculations in this direction, see Appendix E.

Question 4. Can the arithmetic properties of the sequences considered by Lehmer be developed in the same way that the arithmetic properties of binary sequences have? See Ribenboim [Rib95a], Stewart [Ste77] for background, and van der Poorten [Poo89] and references therein for an introduction to the large body of results on recurrence sequences in general.

1.7 When Does the Measure Vanish?

Lehmer's problem asks about small positive values of $m(F)$. In this section we show that the situation where $m(F) = 0$ can be completely understood using Kronecker's lemma.

Theorem 1.31. [KRONECKER] *Suppose that $\alpha \neq 0$ is an algebraic integer and the algebraic conjugates $\alpha_1 = \alpha, \ldots, \alpha_d$ of α all have modulus $|\alpha_j| \leq 1$. Then α is a root of unity.*

Proof. Consider the polynomial

$$F_n(x) = \prod_{i=1}^{d}(x - \alpha_i^n), \tag{1.29}$$

where F_1 is the minimal polynomial for α. The coefficients of F_n are symmetric functions in the algebraic integers α_j^n so they are (rational) integers.

Each of the coefficients is uniformly bounded as n varies because $|\alpha_j| \leq 1$ for all j, so the set

$$\{F_n\}_{n \in \mathbb{N}}$$

must be finite. It follows that there is a pair $n_1 \neq n_2$ for which

$$F_{n_1} = F_{n_2},$$

so

$$\{\alpha_1^{n_1}, \ldots, \alpha_d^{n_1}\} = \{\alpha_1^{n_2}, \ldots, \alpha_d^{n_2}\}.$$

For each permutation $\tau \in S_d$ (the permutation group on d symbols), define an action of τ on the set of roots by

$$\alpha_i^{n_1} = \alpha_{\tau(i)}^{n_2}.$$

Then if τ has order r in S_d,

$$\alpha_i^{n_1^r} = \alpha_i^{n_2^r},$$

so

$$\alpha_i^{n_1^r}\left(\alpha_i^{n_2^r - n_1^r} - 1\right) = 0,$$

which shows that α_i must be a unit root since $\alpha_i \neq 0$.

Remark 1.32. Kronecker's lemma relates an analytic property of algebraic numbers (a condition on the modulus of the zeros) to an algebraic property (that the zeros must be torsion points in the group of complex numbers of modulus one).

A polynomial in $\mathbb{Z}[x]$ is called *primitive* if the coefficients have no non-trivial common factor.

Theorem 1.33. *Suppose $F \in \mathbb{Z}[x]$ is non-zero, primitive and $F(0) \neq 0$. Then $m(F) = 0$ if and only if all the zeros of F are roots of unity.*

Proof. Assume that all the zeros of F are roots of unity. Then the leading coefficient of F must be ± 1 since F divides $x^N - 1$ for some $N \geq 1$. So, from the definition, $m(F) = 0$.

Conversely, if $m(F) = 0$ then it is clear that F must be (plus or minus) a monic polynomial, so all the zeros are algebraic integers, and all must have modulus less than or equal to 1. Apply Kronecker's lemma to see they must all be unit roots.

Remark 1.34. We could restate this by saying that for primitive F, $m(F) = 0$ if and only if F is a monomial times a cyclotomic polynomial.

Exercise 1.16. If $F \in \mathbb{Z}[x]$ is cyclotomic, prove that $F^* = \pm F$.

2. Dynamical Systems

The first connection between Mahler's measure and algebraic dynamical systems is that the *topological entropy* of such systems always turns out to be the Mahler measure of an integral polynomial.

The expression $\sum \log^+ | \cdot |$ arose first in this context as the measure-theoretic entropy of an automorphism of the 2-torus in the work of Sinai [Sin59]. Abramov [Abr59] showed that the measure-theoretic entropy of the solenoidal automorphism determined by a rational $\frac{a}{b}$ in lowest terms was given by $\log \max\{|a|, |b|\}$, which coincides with the Mahler measure of $bx - a$. Yuzvinskii [Yuz67] finally obtained a formula for automorphisms of finite-dimensional solenoids, and his formula is again a Mahler measure.

Berg [Ber69] showed that in this setting the measure-theoretic entropy and the topological entropy coincide, and later Bowen [Bow71] found a general framework for computing the topological entropy. A very simple proof of Yuzvinskii's formula using Bowen's approach and the adele ring is in Lind and Ward [LW88].

In this chapter the toral case is described first, using only methods of elementary Fourier analysis. The solenoid case is presented following Schmidt's monograph [Sch95] and uses more sophisticated methods. For the toral case some knowledge of Lebesgue measure is needed (see Rudin [Rud74, Section 2.19]), and for the general case Haar measure is needed (see Rudin [Rud62, Section 1.1] or Walters [Wal82, Section 0.6] for an introduction and Hewitt and Ross [HR63] for complete details).

2.1 Dynamical Interpretation: Toral Case

In this section we introduce a simple family of dynamical systems and relate various dynamical properties to quantities that appear in Chapter 1. To fully understand these, some ideas from abelian harmonic analysis are needed. In the next section we will give a more systematic construction of a family of dynamical systems parametrized by integer polynomials.

For our purposes a (topological) *dynamical system* is a continuous map $T : X \to X$ of a compact metric space X; if X is a metrizable compact group and T a homomorphism, then it is an *algebraic* dynamical system.

Let $\mathbb{T}^d = \mathbb{R}^d/\mathbb{Z}^d$ denote the additive d-dimensional torus. This is a compact abelian group, and any continuous surjective endomorphism $T :$ $\mathbb{T}^d \to \mathbb{T}^d$ is of the form

$$T(x) = T_A(x),$$

where A is a non-singular $d \times d$ integer matrix, and T_A is defined by

$$T_A \begin{bmatrix} x_1 \\ \vdots \\ x_d \end{bmatrix} = A \begin{bmatrix} x_1 \\ \vdots \\ x_d \end{bmatrix} \bmod 1,$$

(see Exercise 2.1). The map T_A is surjective and therefore preserves the Lebesgue measure λ on $\mathbb{T}^d$ (cf. Walters [Wal82, Section 0.6]).

Exercise 2.1. [1] Prove that a closed subgroup of $\mathbb{T}$ must be either a finite group $\{\frac{j}{k}\}_{j=1,\dots,k}$ for some $k \geq 1$ or all of $\mathbb{T}$.
[2] Prove that the only continuous automorphisms of $\mathbb{T}$ are the identity and $x \mapsto -x$.
[3] Prove that the only continuous endomorphisms of $\mathbb{T}$ are the maps $x \mapsto nx$ for some $n \neq 0$.
[4] By considering the projections $\mathbb{T}^d \to \mathbb{T}$ onto each coordinate, prove that any surjective continuous endomorphism of $\mathbb{T}^d$ must be given by a non-singular integer matrix.

Definition 2.1. The measure-preserving transformation T_A is said to be *ergodic* if every square-integrable function $f : \mathbb{T}^d \to \mathbb{C}$ with the property that

$$f(T_A x) = f(x) \text{ a.e.}$$

is almost everywhere equal to a constant.

Let the eigenvalues of the matrix A be $\{\alpha_1, \dots, \alpha_d\}$.

Lemma 2.2. *The transformation T_A is ergodic if and only if no eigenvalue α_i of A is a root of unity.*

Proof. The set of functions

$$\chi_{(k_1,\dots,k_d)}(t_1, \dots, t_d) = \chi_{\mathbf{k}}(\mathbf{t}) = e^{2\pi i(k_1 t_1 + \dots + k_d t_d)} = e^{2\pi i(\mathbf{k} \cdot \mathbf{t})}$$

for $\mathbf{k} \in \mathbb{Z}^d$ forms an orthonormal basis for the space $L^2(\mathbb{T}^d)$ of square-integrable complex-valued functions on $\mathbb{T}^d$. The action of T_A on this collection of functions is given by

$$\chi_{\mathbf{k}} \circ T_A = \chi_{A_t \mathbf{k}} \tag{2.1}$$

where A_t is the transpose of A.

Assume first that A has a root of unity as an eigenvalue. Then for some $p > 0$, $A_t^p - I$ is a singular linear map from $\mathbb{Q}^d$ to $\mathbb{Q}^d$. It follows that there is a rational vector $\mathbf{v} \in \mathbb{Q}^d \setminus \{0\}$ with

$$(A_t^p - I)\mathbf{v} = 0. \tag{2.2}$$

By clearing fractions in (2.2), we may find an integer vector $\mathbf{w} \in \mathbb{Z}^d \setminus \{0\}$ with

$$(A_t^p - I)\mathbf{w} = 0. \tag{2.3}$$

This means that

$$\chi_\mathbf{w} \circ T_A^p = \chi_\mathbf{w}. \tag{2.4}$$

Assume p has been chosen minimally subject to (2.4). Then the function

$$f = \chi_\mathbf{w} + \chi_\mathbf{w} \circ T_A + \ldots + \chi_\mathbf{w} \circ T_A^{p-1}$$

is invariant under T_A and, by minimality of p, is a sum of distinct functions of the form $\chi_\mathbf{k}$. Since the family $\{\chi_\mathbf{k}\}_{\mathbf{k} \in \mathbb{Z}^d}$ is orthonormal, this means that f is non-constant, so T_A is non-ergodic.

Conversely, suppose T_A is not ergodic. Then there is a non-constant function $f \in L^2(\mathbb{T}^d)$ such that $f \circ T_A = f$. Enumerate the countable set $\{\chi_\mathbf{k}\}_{\mathbf{k} \in \mathbb{Z}^n}$ as $\{\phi_0, \phi_1, \phi_2, \ldots\}$ where $\phi_0 = \chi_{(0,\ldots,0)}$ is the constant function 1. By Fourier theory for square-integrable functions, the Fourier series of f converges in $L^2(\mathbb{T}^d)$ to f:

$$f = \sum_{j=0}^{\infty} a_j \phi_j$$

where $a_j = \int_{\mathbb{T}^d} f(x)\overline{\phi_j(x)} d\lambda(x)$ is the Fourier coefficient corresponding to ϕ_j. By Parseval's formula,

$$\sum_{j=0}^{\infty} |a_j|^2 = \|f\|_2^2 := \int_{\mathbb{T}^d} |f|^2 < \infty. \tag{2.5}$$

A different form of Parseval formula appeared in Section 1.4 for holomorphic functions (see Lemma 1.20); equation (2.5) holds simply because $\{\chi_\mathbf{k}\}$ is an orthonormal basis in the Hilbert space $L^2(\mathbb{T}^d)$ (see Katznelson [Kat76, Lemma 5.4]).

Since f is non-constant one of the other coefficients (a_s with $s \neq 0$ say), must be non-zero. Since f is invariant under T_A and Fourier coefficients are unique, the coefficient of the characters

$$\phi_s, \phi_s \circ T_A, \phi_s \circ T_A^2, \ldots$$

must all be equal to $a_s \neq 0$. By the summability condition (2.5), these cannot all be distinct members of the set $\{\phi_0, \phi_1, \ldots\}$ so $\phi_s \circ A^p = \phi_s \circ A^q$ for some

$p > q \geq 0$. The matrix A is non-singular, so acts injectively on the set of ϕ_j's; it follows that

$$\phi_s \circ A^{p-q} = \phi_s.$$

Using the description of the action in (2.1) we deduce that there is an integer vector $\mathbf{k} \in \mathbb{Z}^d \backslash \{0\}$ for which

$$\mathbf{k} = A_t^{p-q} \mathbf{k},$$

so A_t^{p-q} has 1 as an eigenvalue, which requires A to have a $(p-q)$th root of unity as an eigenvalue.

The set of points with period n under T_A is defined to be

$$\mathrm{Per}_n(T_A) = \{x \in \mathbb{T}^n : T_A^n(x) = x\}.$$

Notice that this is a closed subgroup of $\mathbb{T}^n$, and that if $m|n$ (m divides n) then $\mathrm{Per}_m(T_A)$ is a subgroup of $\mathrm{Per}_n(T_A)$.

Exercise 2.2. Let $T_A : \mathbb{T}^d \to \mathbb{T}^d$ be an automorphism of the torus. Show that every point of the form $(a_1, \ldots, a_d) + \mathbb{Z}^d$ with $a_i \in \mathbb{Q}$ for all i is periodic under T_A. If T_A is ergodic, show that no other points are periodic points.

Lemma 2.3. *The number of points of period n under T_A is given by*

$$|\mathrm{Per}_n(T_A)| = |\det(A^n - I)| = |\Delta_n(\chi_A)|$$

if T_A is ergodic, where χ_A is the characteristic polynomial of A.

Before proving this, we will consider a simple example to explain how the expressions arise. Let $\pi : \mathbb{R}^d \to \mathbb{T}^d \cong \mathbb{R}^d / \mathbb{Z}^d$ be the canonical quotient map, with chosen fundamental domain $\mathcal{F} = [0, 1)^d$. The number of points with period n under T_A is the cardinality of $\ker(T_A^n - I)$, and this in turn is equal to the number of solutions to the equation

$$(A^n - I)\mathbf{x} \in \mathbb{Z}^d; \quad \mathbf{x} \in \mathcal{F}.$$

Equivalently, it is the cardinality of the set

$$(A^n - I)^{-1}(\mathbb{Z}^d) \cap \mathcal{F}. \tag{2.6}$$

This gives a geometric proof. We also give a more sophisticated proof using harmonic analysis.

Example 2.4. Let $A = \begin{bmatrix} 4 & 3 \\ 1 & 1 \end{bmatrix}$. Then we may compute the number of fixed points for T_A using (2.6): $(A - I)^{-1} = \begin{bmatrix} 0 & 1 \\ 1/3 & -1 \end{bmatrix}$, so

$$(A - I)^{-1} \begin{bmatrix} a \\ b \end{bmatrix} = \begin{bmatrix} b \\ (1/3)a - b \end{bmatrix}$$

with $a, b \in \mathbb{Z}$. This falls in $[0, 1)^2$ if and only if $b = 0$ and $a = 0, 1$, or 2. We deduce that

$$|\text{Per}_1(T_A)| = 3,$$

and by applying the projection map π to each point found, we see that

$$\text{Per}_1(T_A) = \left\{ \begin{bmatrix} 0 \\ 0 \end{bmatrix}, \begin{bmatrix} 0 \\ 1/3 \end{bmatrix}, \begin{bmatrix} 0 \\ 2/3 \end{bmatrix} \right\}.$$

Notice that Lemma 2.3 shows that T_A is ergodic if and only if $\text{Per}_n(T_A)$ is finite for every $n \in \mathbb{N}_{>0}$, and in that case

$$|\text{Per}_n(T_A)| = |\det(A^n - I)| = \prod_{i=1}^{d} |\alpha_i^n - 1|.$$

Lemma 2.3 gives another explanation of why the quantity $\Delta_n(F)$ in (1.1) is an integer – it is the cardinality of a finite group – and shows that the sequence $(\Delta_n(F))$ is a divisibility sequence (cf. Exercise 1.2) in that

$$m | n \Rightarrow |\Delta_m(F)| \, \big| \, |\Delta_n(F)|.$$

Proof (of Lemma 2.3 via abelian groups). It is enough to show that if B is a non-singular integer matrix, then

$$B^{-1}(\mathbb{Z}^d) \cap F \text{ has cardinality } |\det(B)|, \tag{2.7}$$

where $F = [0, 1)^d$ is as usual the chosen fundamental domain for the quotient map $\pi : \mathbb{R}^d \to \mathbb{T}^d$. First notice that

$$\mathbf{x} \in B^{-1}(\mathbb{Z}^d) \cap F \text{ if and only if } B\mathbf{x} \in \mathbb{Z}^d \cap BF, \tag{2.8}$$

so it is enough to show that the volume $|\det(B)|$ of the half-open d-dimensional parallelepiped BF is exactly equal to the number of integer lattice points that lie in it:

$$|BF \cap \mathbb{Z}^d| = |\det(B)|. \tag{2.9}$$

For $d = 2$ this is simply Pick's theorem. In the general case, we may argue as follows. The set BF is a fundamental domain for the subgroup $B\mathbb{Z}^d$ in $\mathbb{R}^d$. So integer points in BF correspond one-to-one to cosets of $B\mathbb{Z}^d$ in $\mathbb{Z}^d$. Thus $|BF \cap \mathbb{Z}^d|$ is equal to the cardinality of the finite abelian group $\mathbb{Z}^d / B\mathbb{Z}^d$, which is $|\det(B)|$.

Exercise 2.3. Give a geometric proof of the equality (2.9).

Proof (of Lemma 2.3 via harmonic analysis). Let $\theta : G \to G$ be a continuous homomorphism with closed image on some locally compact abelian group G. The dual map is a homomorphism $\hat{\theta} : \hat{G} \to \hat{G}$. The kernel of θ is a closed subgroup of G, and general results from duality theory show that

$$\widehat{\ker \theta} \cong \operatorname{coker} \hat{\theta}. \tag{2.10}$$

In our situation, let $\theta = T_A^n - I$, $G = \mathbb{T}^d$; then the result (2.10) shows that the dual group of $\operatorname{Per}_n(T_A)$ is

$$\ker \widehat{(T_A^n - I)} \cong \operatorname{coker} (\widehat{T_A^n - I}) = \mathbb{Z}^d / (A^n - I)\mathbb{Z}^d.$$

If T_A is ergodic, it follows from Lemma 2.2 that $A^n - I$ is non-singular for all $n \geq 1$, so $\operatorname{Per}_n(T_A)$ is finite for all $n \geq 1$. The last assertion now follows from the fact that any finite abelian group is isomorphic to its dual group.

2.2 Topological Entropy

The topological entropy of a continuous map $T : X \to X$, where X is a compact Hausdorff space was introduced by Adler, Konheim and McAndrew [AKM65] as a measure of topological complexity analogous to a measure-theoretic notion introduced by Kolmogorov and Sinai. An *open cover* $\mathcal{U}$ of X is a collection of open subsets of X whose union is all of X, and a *subcover* is a subset of $\mathcal{U}$ that still covers all of X. Write $|\mathcal{U}|$ for the cardinality of the set $\mathcal{U}$. The *join* of two open covers $\mathcal{V}$ and $\mathcal{U}$, written $\mathcal{V} \vee \mathcal{U}$, is the open cover comprising all the non-empty members of $\{V \cap U : V \in \mathcal{V}, U \in \mathcal{U}\}$. For any open cover $\mathcal{U}$, let

$$N(\mathcal{U}) = \min\{|\mathcal{V}| : \mathcal{V} \text{ is a subcover of } \mathcal{U}\};$$

by compactness this is always finite.

Definition 2.5. The *topological entropy* of T is defined by

$$h_{top}(T) = \sup_{\mathcal{U}} \lim_{n \to \infty} \frac{1}{n} \log N\left(\mathcal{U} \vee T^{-1}(\mathcal{U}) \vee \ldots \vee T^{-(n-1)}(\mathcal{U})\right)$$

where the supremum is taken over all open covers.

Of course the existence of the limit needs to be shown – see Adler, Konheim and McAndrew [AKM65] or Petersen [Pet83, Proposition 3.2].

For our purposes – where we are dealing with algebraic dynamical systems – it is simpler to use a different definition of topological entropy. Introduce a natural metric ρ on $\mathbb{T}^d$, by setting

$$\rho(x + \mathbb{Z}^d, y + \mathbb{Z}^d) = \min_{z, w \in \mathbb{R}^d : z - w = x - y} \{\|z - w\|\} \tag{2.11}$$

where $\| \cdot \|$ is the usual Euclidean norm on $\mathbb{R}^d$.

The *topological entropy* of the toral endomorphism T_A is defined to be

$$h(T_A) = \lim_{\epsilon \searrow 0} \lim_{n \to \infty} -\frac{1}{n} \log \lambda \left(\bigcap_{j=0}^{n-1} T_A^{-j}(B_\epsilon) \right), \qquad (2.12)$$

where B_ϵ is the open ball around the identity of radius ϵ with respect to the metric ρ. The entropy is thus a measure of volume growth under the transformation; it also is a measure of orbit complexity of the dynamical system. The name 'topological' is attached to this quantity because *for endomorphisms of compact metric groups* it coincides with the topological entropy defined using purely topological notions (counting growth in open covers, cf. Definition 2.5) or via metric space notions (spanning and separating sets cf. Bowen [Bow71]); the equality is due to Bowen. The expression (2.12) may also be used to define the topological entropy of any locally compact metric group endomorphism, with λ meaning Haar measure on the group, and using any translation-invariant metric compatible with the topology.

Theorem 2.6. *The topological entropy of the transformation T_A is equal to the Mahler measure of the characteristic polynomial of the matrix A.*

This is proved in Walters [Wal82, Section 8.4]. An example is discussed below to illustrate how the eigenvalues of the matrix arise in the entropy calculation, and we prove a generalization of a special case below in Theorem 2.12. The special case is that we assume the matrix is (conjugate to) the companion matrix of its characteristic polynomial; the generalization is that we allow the entries in the matrix to become rationals.

Definition 2.7. Let $F(x) = x^d + a_{d-1}x^{d-1} + \ldots + a_0$ be a polynomial. The *companion matrix* to F is the matrix

$$\begin{bmatrix} 0 & 1 & 0 & \cdots & 0 \\ 0 & 0 & 1 & \cdots & 0 \\ \vdots & \vdots & \vdots & \ddots & \vdots \\ 0 & 0 & 0 & \cdots & 1 \\ -a_0 & -a_1 & -a_2 & \cdots & -a_{d-1} \end{bmatrix}.$$

Notice that the characteristic equation of the companion matrix is $F(x) = 0$, so the companion matrix allows us to exhibit an endomorphism of the torus with given characteristic equation very easily.

Theorem 2.6 relates the possible entropies of toral automorphisms directly to Lehmer's problem (cf. Remark 1.7[3]). In fact Lind [Lin74, Theorem 2] has shown that Lehmer's problem is equivalent to the problem of the existence of an ergodic automorphism of the infinite torus $\mathbb{T}^N$ with finite entropy. On the general problem of determining the set of possible entropies of compact group automorphisms, Lind [Lin77, Theorem 9.3] has shown that this set is either a countable subset of $[0, \infty]$ whose closure omits 0 or is all of $[0, \infty]$, depending on the solution to Lehmer's problem. The measure-theoretic structure of ergodic automorphisms of compact groups is completely understood: Lind [Lin77] has shown that they are measurably isomorphic to *Bernoulli shifts* (see Petersen [Pet83, Sections 6.4 and 6.5]). It follows that Lehmer's problem is equivalent to determining if there is an ergodic group automorphism which is measurably the same as any Bernoulli shift, or whether only Bernoulli shifts with special entropies have such a model.

Corollary 2.8. *If T_A is an ergodic toral endomorphism, then the growth rate of periodic points exists and equals the topological entropy:*

$$\lim_{n \to \infty} \frac{1}{n} \log |\mathrm{Per}_n(T_A)| = h(T_A). \tag{2.13}$$

Proof. This follows at once from Lemma 1.10, Lemma 2.3 and Theorem 2.6.

Finally, assume that the transformation T_A is an automorphism (that is, $\det(A)$ is ± 1). Then the transformation T_A is said to be *expansive* if there is a constant $\delta > 0$ with the property that for any pair $x \neq y$ in $\mathbb{T}^d$, there is an $n \in \mathbb{Z}$ with the property that

$$\rho(T_A^n x, T_A^n y) > \delta.$$

Lemma 2.9. *The invertible transformation T_A is expansive if and only if no eigenvalue of A has unit modulus.*

This well-known result may be seen directly using the Jordan form of the complexified matrix (outlined in Walters [Wal82, Chapter 8]) or as part of the general theory of hyperbolic dynamical systems (see Katok and Hasselblatt [KH95]). Eisenberg in [Eis66] proved the analogous statement for matrices acting on topological vector spaces (which for real vector spaces implies the toral result) over any non-discrete topological field.

As with Theorem 2.6, we prove this for matrices in companion form to their characteristic polynomial (and allow rational entries) below in Theorem 2.13. A matrix is called *hyperbolic* if no eigenvalue has modulus one. To give an indication of how the hyperbolicity of the matrix forces the map to be expansive a special case is proved here.

Proof (of special case in Lemma 2.9). Consider the toral automorphism T_A where $A = \begin{bmatrix} 2 & 1 \\ 1 & 1 \end{bmatrix}$. The eigenvalues are

$$\lambda = \frac{3 + \sqrt{5}}{2} > 1 \quad \text{and} \quad \mu = \frac{3 - \sqrt{5}}{2} < 1.$$

We will show that the automorphism T_A is expansive. The action of T_A on the torus $\mathbb{T}^2$ is illustrated in Figure 2.1: the square denotes the fundamental domain

$$\mathcal{F} = \left\{ \begin{bmatrix} x \\ y \end{bmatrix} : 0 \le x < 1, 0 \le y < 1 \right\},$$

which is identified with $\mathbb{T}^2$ via the restriction of the map $\pi : \mathbb{R}^2 \to \mathbb{R}^2/\mathbb{Z}^2 \cong \mathbb{T}^2$ to the domain $\mathcal{F}$.

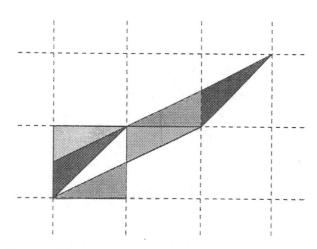

Fig. 2.1. The image of the fundamental domain under T_A

The shading in Figure 2.1 indicates how the different regions in the image of the fundamental domain in $\mathbb{R}^2$ are translated back onto the fundamental domain to describe the map T_A on the quotient $\mathbb{R}^2/\mathbb{Z}^2$.

This matrix is in some sense the simplest (orientation-preserving) hyperbolic matrix, and is therefore a standard example in hyperbolic dynamics (cf. Bowen [Bow78, Chapter 1] and Katok and Hasselblatt [KH95, Section 1.8]).

Since $\rho(T_A^n x, T_A^n y) = \rho(T_A^n(x - y), 0)$, it is enough to show that there is a $\delta > 0$ with the property that any $x \ne 0$ in $\mathbb{T}^2$ has $\rho(T_A^n x, 0) > \delta$ for some $n \in \mathbb{Z}$.

To do this, consider the effect of T_A on points near the identity in $\mathbb{T}^2$ – these correspond to points in the fundamental domain close to any of the four corners, but for convenience these may be viewed as lying close to the origin in $\mathbb{R}^2$ up to translation by an integer vector. If a point $x = \begin{bmatrix} x_1 \\ x_2 \end{bmatrix}$ lies on the *expanding* direction $x_2 = \lambda x_1$ and is within Euclidean distance $\frac{1}{10}$ of the origin, then $T_A(x)$ is obtained by simply scaling by $\lambda > 1$ on the same line. It follows that for some $n > 0$ the point $T_A^n(x)$ is distance at least $\frac{1}{10}$ from the origin. Similarly, if $x = \begin{bmatrix} x_1 \\ x_2 \end{bmatrix}$ lies on the *contracting* direction $x_2 = \mu x_1$, then $T_A^{-1}(x)$ is obtained by scaling x by $\frac{1}{\mu} > 1$ on the same line. It follows that for some $n < 0$ the point $T_A^n(x)$ is distance at least $\frac{1}{10}$ from the origin. In the general case, the point $x = \begin{bmatrix} x_1 \\ x_2 \end{bmatrix}$ lies on neither line. In this case we may draw a rectangle with sides parallel to the expanding and contracting directions, as shown in Figure 2.2.

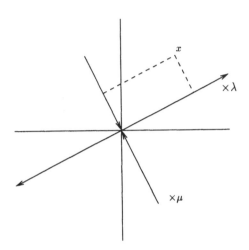

Fig. 2.2. Expanding and contracting directions

The effect of T_A (for x close to the origin) is given by scaling the coordinate of x on the expanding line by λ, and on the contracting line by μ. Thus for some $n > 0$, $T_A^n(x)$ is distance at least $\frac{1}{10}$ from the origin.

It follows that T_A is expansive.

In a similar spirit, we now indicate a special case of Theorem 2.6.

Proof (of special case in Theorem 2.6). Define an adapted metric on a neighbourhood of the identity in the torus by using the maximum metric of the projections onto the expanding and contracting directions. It is clear that we may use this metric to compute the topological entropy using (2.12). An open ball B of radius ϵ in this metric looks like a rectangle with sides parallel to the expanding and contracting directions. The effect of T_A^{-1} on this open ball is to dilate the sides parallel to the expanding directions by a factor of λ^{-1} and shrink the sides parallel to the contracting directions by a factor of μ^{-1}. It follows that for large values of n, $\bigcap_{j=0}^{n-1} T_A^{-j}(B)$ is a very thin rectangle with area $\epsilon^2 \mu^{-(n-1)}$. For large n, $T_A^{-n}(B)$ is the projection of a long thin rectangle in $\mathbb{R}^2$ which only intersects $\bigcap_{j=0}^{n-1} T_A^{-j}(B)$ once, so (2.12) reduces to

$$h(T_A) = \log \lambda.$$

In summary: to each monic integer polynomial F we can associate an endomorphism of the torus by letting the companion matrix of the polynomial act linearly. The zeros of the polynomial correspond to the eigenvalues of the matrix. If the polynomial has no roots of unity as zeros, the corresponding dynamical system is ergodic. Assuming ergodicity, the growth rate of periodic points for this dynamical system converges to the entropy. Assuming expansiveness, the growth measure used by Lehmer also converges to the entropy. The Mahler measure corresponds to the topological entropy of the associated dynamical system.

2.3 Dynamical Interpretation: Solenoid Case

Dynamical properties of the toral endomorphisms of the last section were shown to be associated to properties of the associated monic characteristic polynomial of the matrix defining the endomorphism. Mahler's measure is also defined for non-monic polynomials, and in this section we describe the associated dynamical systems. These are automorphisms of *solenoids* (except for the degenerate case of a constant polynomial), and several new phenomena arise here.

Fix a polynomial $F(x) = a_d x^d + a_{d-1} x^{d-1} + \ldots + a_0 \in \mathbb{Z}[x]$ with $a_d a_0 \neq 0$. Associate to F the compact group X_F defined by

$$X_F = \left\{ z = (z_k) \in \mathbb{T}^{\mathbb{Z}} : a_0 z_k + a_1 z_{k+1} + \ldots + a_d z_{k+d} = 0 \text{ for all } k \in \mathbb{Z} \right\}.$$
$$(2.14)$$

This is a closed subgroup of $\mathbb{T}^{\mathbb{Z}}$, which is compact by Tychonoff's theorem – see Lang [Lan73, Section II.3]; this compact group is also metrizable – an explicit metric is given in (2.16) below. The map $T = T_F : X_F \to X_F$ given by the left shift:

$$(T_F(z))_k = z_{k+1}$$

is a continuous automorphism of the compact group X_F. A basis of open balls around the identity in X_F is given by sets of the form

$$U(N, \epsilon) = \{ z \in X_F : \rho(z_k, 0) < \epsilon \text{ for } |k| = 0, \ldots, N - 1 \} \qquad (2.15)$$

where $\epsilon > 0$, $N \geq 1$ and ρ is the metric on $\mathbb{T}$ defined in (2.11). The same topology is induced by the metric

$$d(z, z') = \sum_{n \in \mathbb{Z}} 2^{-|n|} \rho(z_n, z'_n). \qquad (2.16)$$

Notice that the assumption $a_0 a_d \neq 0$ is not a real restriction for non-zero polynomials: the groups X_F and $X_{Fx^{\pm 1}}$ are identical, so we may always assume that $a_0 a_d \neq 0$.

Exercise 2.4. The case $F = 0$ was omitted. Show that the resulting dynamical system T_F on X_F is the shift map on $\mathbb{T}^{\mathbb{Z}}$, with infinitely many points of period n for any $n \geq 1$ and with infinite entropy.

Example 2.10. [1] If $F(x) = a$, a constant, then

$$X_F = \prod_{k=-\infty}^{\infty} \left\{ 0, \frac{1}{|a|}, \frac{2}{|a|}, \ldots, \frac{|a|-1}{|a|} \right\},$$

and so the map T_A is the left shift on the space of one-sided sequences with $|a|$ symbols. There are $|a|^n$ points of period n, and the topological entropy is easily seen to be $\log |a|$.

[2] If $F(x) = x^2 - 5x + 1$, then the condition appearing in (2.14) is

$$z_k - 5z_{k+1} + z_{k+2} = 0$$

for all $k \in \mathbb{Z}$. Notice that the pair of points (z_0, z_1) determine all the other coordinates of $z = (z_k)$ uniquely, since

$$\begin{bmatrix} z_{k+1} \\ z_{k+2} \end{bmatrix} = \begin{bmatrix} 0 & 1 \\ -1 & 5 \end{bmatrix} \begin{bmatrix} z_k \\ z_{k+1} \end{bmatrix} \quad \text{and} \quad \begin{bmatrix} z_{k-1} \\ z_k \end{bmatrix} = \begin{bmatrix} 5 & -1 \\ 1 & 0 \end{bmatrix} \begin{bmatrix} z_k \\ z_{k+1} \end{bmatrix}$$

for all $k \in \mathbb{Z}$. This shows that the map $\theta : X_F \to \mathbb{T}^2$ defined by $\theta((z_k)) = \begin{bmatrix} z_0 \\ z_1 \end{bmatrix}$ is a group isomorphism that intertwines the shift map T_F and the toral automorphism T_A where $A = \begin{bmatrix} 0 & 1 \\ -1 & 5 \end{bmatrix}$. That is, the following diagram commutes:

$$
\begin{array}{ccc}
(\ldots, z_{-1}, \overset{0}{\overbrace{z_0}}, z_1, \ldots) & \overset{T_F}{\longrightarrow} & (\ldots, z_0, \overset{0}{\overbrace{z_1}}, z_2, \ldots) \\
\theta \downarrow & & \downarrow \theta \\
\begin{bmatrix} z_0 \\ z_1 \end{bmatrix} & \overset{A}{\longrightarrow} & \begin{bmatrix} z_1 \\ z_2 \end{bmatrix}
\end{array}
$$

The topological entropy of the map T_F is $\log \left(\frac{5}{2} + \frac{1}{2}\sqrt{21} \right)$ by Theorem 2.12.

[3] If $F(x) = x - 2$, then the condition appearing in (2.14) is that $2z_k = z_{k+1}$ for all k. It follows that the 'future' $(z_k)_{k>0}$ of an element $z \in X_F$ is completely determined by $z_0 \in \mathbb{T}$, while the 'past' is determined up to choosing one of the two possible solutions to the equation $2z_k = z_{k+1}$ for $k = -1, -2, \ldots$ in turn. It may be shown that in this case X_F is isomorphic to the dual group of the dyadic rationals, $X_F \cong \mathbb{Z}[\frac{1}{2}]$, and the shift map T_F is dual to the automorphism $x \mapsto 2x$ on $\mathbb{Z}[\frac{1}{2}]$. Here the entropy is $\log 2$.

[4] More generally, if $F \in \mathbb{Z}[x]$ is monic, then a similar argument shows that (X_F, T_F) is algebraically isomorphic to the endomorphism

$$T_A : \mathbb{Z}^d\widehat{[A, A^{-1}]} \to \mathbb{Z}^d\widehat{[A, A^{-1}]}$$

dual to $x \mapsto A_t x$, where A is the companion matrix of F, and d is the degree of F.

[5] If $F(x) = 2x - 1$, then it is clear that the group X_F is isomorphic to the group obtained in [3], and under this isomorphism the shift map is sent to the inverse of the doubling map in [3].

Exercise 2.5. Imitate the proof of Lemma 2.2 to show that T_F is ergodic if and only no zero of F is a root of unity.

The topological entropy of T_F turns out to be the Mahler measure $m(F)$. We prove this following Schmidt [Sch95, Section 17]. An alternative approach using the notion of adeles may be found in Lind and Ward [LW88]. First we need a continuity result that relates the volume growth of iterated matrices to the eigenvalues. Let λ_X denote Haar measure on the group X (for our purposes X will be $\mathbb{C}^* \simeq \mathbb{R}^{2k}$, so this is just Lebesgue measure). Write $\|\cdot\|_\infty$ for the max metric on $\mathbb{C}^*$.

Lemma 2.11. *Let A be a $k \times k$ complex matrix, with eigenvalues $\{\alpha_1, \ldots, \alpha_k\}$. Define*

$$\bar{h}_m(A) = -\frac{1}{m} \log \lambda_{\mathbb{C}^k} \left(\bigcap_{j=0}^{m-1} A^{-j} \left(\{ \mathbf{z} \in \mathbb{C}^* : \|\mathbf{z}\|_\infty < 1 \} \right) \right).$$

Then

$$\bar{h}_m(A) \longrightarrow \sum_{j=1}^{k} \log^+ |\alpha_j|^2$$

as $m \to \infty$.

Proof. This result may be found in Bowen's paper [Bow71]. Our proof follows that of Walters [Wal82, Theorem 8.14].

Let $U \simeq \mathbb{C}^{p_1}$ be the subspace of $\mathbb{C}^*$ spanned by all eigenvectors of A with modulus greater than 1 and let $S \simeq \mathbb{C}^{p_2}$ be the subspace of $\mathbb{C}^*$ spanned by all eigenvectors of A with modulus less than or equal to 1. Then $\mathbb{C}^* = U \oplus S$. Write $A_U = A|_U$ and $A_S = A|_S$ for the restrictions to these invariant subspaces. Since all norms on $\mathbb{C}^*$ are equivalent, we may use the norm $\max\{\|\cdot\|_U, \|\cdot\|_S\}$. Then

$$\bar{h}_m(A) = -\frac{1}{m} \log \underbrace{\lambda_U \left(\bigcap_{j=0}^{m-1} A_U^{-j} \left(\{ \mathbf{z} \in \mathbb{C}^{p_1} : \|\mathbf{z}\|_U < 1 \} \right) \right)}_{A_U(m)}$$

$$-\frac{1}{m} \log \underbrace{\lambda_S \left(\bigcap_{j=0}^{m-1} A_S^{-j} \left(\{ \mathbf{z} \in \mathbb{C}^{p_2} : \|\mathbf{z}\|_S < 1 \} \right) \right)}_{A_S(m)}.$$

Now

$$\lambda_U \left(A_U(m) \right) \leq \lambda_U \left(A_U^{-(m-1)} \left(\{ \mathbf{z} \in \mathbb{C}^{p_1} : \|\mathbf{z}\|_U < 1 \} \right) \right)$$
$$= |\det(A_U)|^{-2(m-1)} \lambda_U \left(\{ \mathbf{z} \in \mathbb{C}^{p_1} : \|\mathbf{z}\|_U < 1 \} \right), \text{ and}$$
$$\lambda_S \left(A_S(m) \right) \leq \lambda_S \left(\{ \mathbf{z} \in \mathbb{C}^{p_2} : \|\mathbf{z}\|_S < 1 \} \right).$$

It follows that

$$\bar{h}_m(A) \geq 2 \left(\frac{m-1}{m} \right) \log |\det A_U|$$
$$- \frac{1}{m} \log \lambda_S \left(\{ \mathbf{z} \in \mathbb{C}^{p_2} : \|\mathbf{z}\|_S < 1 \} \right)$$
$$- \frac{1}{m} \log \lambda_U \left(\{ \mathbf{z} \in \mathbb{C}^{p_1} : \|\mathbf{z}\|_U < 1 \} \right),$$

so

$$\liminf_{m \to \infty} \bar{h}_m(A) \geq \sum_{j=1}^{k} \log^+ |\alpha_j|^2. \tag{2.17}$$

For the reverse inequality, use the Jordan normal form for A to write

$$\mathbb{C}^* = V_1 \oplus V_2 \oplus \ldots \oplus V_s$$

with $A_i = A|_{V_i}$ having all eigenvalues of fixed modulus $r_i > 0$. Write λ_i for Lebesgue measure on V_i. It is enough to show that

$$\limsup_{m \to \infty} \bar{h}_m(A_i) \leq 2 \dim(V_i) \log r_i \text{ if } r_i \geq 1 \text{ and } = 0 \text{ if } r_i < 1. \tag{2.18}$$

Fix $\delta > 0$, $i \in \{1, \ldots, s\}$, and define a function on V_i by

$$\|x\|' = \sum_{n=0}^{\infty} \frac{\|A_i^n x\|}{(r_i + \delta)^n}. \tag{2.19}$$

Since

$$\left(\frac{\|A_i^n x\|}{(r_i + \delta)^n} \right)^{1/n} \leq \frac{\|A_i^n\|^{1/n} \cdot \|x\|^{1/n}}{(r_i + \delta)} \longrightarrow \frac{r_i}{r_i + \delta} < 1,$$

the series (2.19) converges and so clearly defines a norm. Also,

$$\|x\| \leq \|x\|' \leq \|x\| \cdot \underbrace{\sum_{n=0}^{\infty} \frac{\|A_i^n\|}{(r_i + \delta)^n}}_{C}$$

so that $\|\cdot\|$ and $\|\cdot\|'$ are equivalent norms. It is therefore enough to consider

$$- \frac{1}{m} \log \lambda_{V_i} \left(\bigcap_{j=0}^{m-1} A_i^{-j} \left(\{ \mathbf{z} \in V_i : \|\mathbf{z}\|' < 1 \} \right) \right).$$

Now
$$A_i^{-j}\left(\{z \in V_i : \|z\|' < 1\}\right) \supset \{z \in V_i : \|z\|' < (r_i + \delta)^{-j}\}$$
by construction. So for $r_i + \delta \geq 1$

$$\bigcap_{j=0}^{m-1} A_i^{-j}\left(\{z \in V_i : \|z\|' < 1\}\right) \supset \{z \in V_i : \|z\|' < (r_i + \delta)^{-(m-1)}\}$$

$$\supset \{z \in V_i : \|z\| < C(r_i + \delta)^{-(m-1)}\},$$

and for $r_i + \delta < 1$

$$\bigcap_{j=0}^{m-1} A_i^{-j}\left(\{z \in V_i : \|z\|' < 1\}\right) \supset \{z \in V_i : \|z\|' < 1\}$$

$$\supset \{z \in V_i : \|z\| < C\}.$$

It follows that for $r_i + \delta \geq 1$,

$$\lambda_{V_i}\left(\bigcap_{j=0}^{m-1} A_i^{-j}\left(\{z \in V_i : \|z\|' < 1\}\right)\right)$$

$$\geq \frac{1}{(r_i + \delta)^{2 \dim V_i (m-1)}} \lambda_{V_i}\left(\{z \in V_i : \|z\| < C\}\right),$$

and for $r_i + \delta < 1$,

$$\lambda_{V_i}\left(\bigcap_{j=0}^{m-1} A_i^{-j}\left(\{z \in V_i : \|z\|' < 1\}\right)\right) \geq \lambda_{V_i}\left(\{z \in V_i : \|z\| < C\}\right).$$

This proves (2.18) and concludes the proof of Lemma 2.11.

Theorem 2.12. *Let $F \in \mathbb{Z}[x]$ be a polynomial. Then the topological entropy of T_F on X_F is $m(F)$.*

Proof. We follow Schmidt [Sch95, Proposition 17.2] for this proof. If $F = ax^m \neq 0$ is a monomial, then the compact group X_F is isomorphic to $\prod_{\mathbb{Z}} \{0, 1, \ldots, |a|\}$, and the map T_F is the left shift. A ball of radius ϵ about the origin in X_F looks like a 'cylinder' set

$$B_\epsilon = \{z = (z_k) : z_k = 0 \text{ for } |k| \leq R(\epsilon)\}$$

for some $R(\epsilon) \to \infty$ as $\epsilon \to 0$. The measure of this set is $\lambda(B_\epsilon) = \left(\frac{1}{|a|}\right)^{2R(\epsilon)+1}$.
On the other hand,

$$\bigcap_{j=0}^{n-1} T_F^{-j}(B_\epsilon) = \{z = (z_k) : z_k = 0 \text{ for } -R(\epsilon) \leq k \leq R(\epsilon) + n - 1\},$$

with measure

$$\lambda \left(\bigcap_{j=0}^{n-1} T_F^{-j}(B_\epsilon) \right) = \left(\frac{1}{|a|} \right)^{2R(\epsilon)+n},$$

which gives

$$h(T_F) = \log |a| = m(F)$$

by (2.12).

If $F(x) = 0$, then $h(T_F) = m(F)$ by definition (cf. Exercise 2.4, discussion after Definition 1.4) .

Now assume that F is non-monomial, and write

$$F(x) = a_d x^d + a_{d-1} x^{d-1} + \ldots + a_0 \in \mathbb{Z}[x]$$

with $a_d a_0 \neq 0$, and put $a'_j = a_j / a_d$ for $j = 0, \ldots, d-1$ (we can multiply F by a monomial to ensure it has a non-zero constant term without changing X_F). If $\rho(x, 0) < \frac{1}{2}$ then there is a unique $y \in \mathbb{R}$ with the property that $|y| < \frac{1}{2}$ and $y \bmod 1 = x$ (that is, $(-\frac{1}{2}, \frac{1}{2})$ meets each coset of $\mathbb{Z}$ in $\mathbb{R}$ no more than once). Extending this, we have that for $\epsilon \in (0, \frac{1}{2})$, to any point

$$z \in B(N, \epsilon) = \{ z \in X_F : \rho(z_k, 0) < \epsilon \text{ for } k = 0, \ldots, N-1 \}$$

there corresponds a unique point $\mathbf{y} = (y_0, \ldots, y_{N-1}) = \phi_N(z) \in \mathbb{R}^N$ such that

$$|y_j| < \epsilon \text{ for } j = 0, \ldots, N-1$$

and

$$y_j \bmod 1 = x_j \text{ for } j = 0, \ldots, N-1.$$

Notice that this defines a map ϕ_N on points in X_F that are sufficiently close to the identity.

We want to describe the values of y_j as j changes using a method from recurrence relations. In order to keep the unique representative in $\mathbb{R}$ of points in the circle, we must force them to be close enough to the origin to divide by a_d and multiply by the companion matrix of the monic polynomial $\frac{1}{a_d} F(x)$. Explicitly, if

$$\epsilon < \frac{1}{10|a_d| \times \max_{1 \leq j \leq d} \{ |\alpha_j| + 1 \}},$$

then for $N > d$

$$\phi_N(B(N, \epsilon)) = \{ \mathbf{y} \in \mathbb{R}^N : a'_0 y_k + \ldots + a'_{d-1} y_{k+d-1} + y_{k+d} = 0 \\ \text{for } k = 0, \ldots, N-d-1 \}.$$

Fix a point $\mathbf{y} = (y_0, \ldots, y_{N-1}) \in \phi_N(B(N, \epsilon))$ and set

$$\mathbf{x}_k = (y_k, \ldots, y_{k+d-1}) \in \mathbb{R}^d.$$

Then $\mathbf{x}_{k+1}^t = A \mathbf{x}_k^t$ for $k = 0, \ldots, N-d-2$ where A is the matrix

$$A = \begin{bmatrix} 0 & 1 & 0 & \cdots & 0 \\ 0 & 0 & 1 & \cdots & 0 \\ \vdots & \vdots & \vdots & \ddots & \vdots \\ 0 & 0 & 0 & \cdots & 1 \\ -a'_0 & -a'_1 & -a'_2 & \cdots & -a'_{d-1} \end{bmatrix}.$$

It follows that

$$\phi_d(B(N, \epsilon)) = \{\mathbf{v} \in \mathbb{R}^d : \|A^k \mathbf{v}\|_\infty < \epsilon \text{ for } k = 0, \ldots, N - d - 1\},$$

where $\| \cdot \|_\infty$ as before denotes the max norm on $\mathbb{R}^d$ or $\mathbb{C}^d = \mathbb{R}^{2d}$.

As before, write λ_X for Haar measure on X. Then

$$\frac{\lambda_{X_F}(B(N-1, \epsilon))}{\lambda_{X_F}(B(N, \epsilon))} = |a_d|^{-1} \frac{\lambda_{\mathbb{R}^d}(\phi_d(B(N-1, \epsilon)))}{\lambda_{\mathbb{R}^d}(\phi_d(B(N, \epsilon)))}. \tag{2.20}$$

The coordinates in positions $0, 1, \ldots, d-1$ may be chosen freely, so

$$\lambda_{X_F}(B(d, \epsilon)) = \lambda_{\mathbb{R}^d}(\phi_d(B(d, \epsilon))) = (2\epsilon)^d.$$

It follows that

$$\frac{\lambda_{X_F}(B(d+1, \epsilon))}{\lambda_{\mathbb{R}^d}(\phi_d(B(d+1, \epsilon)))} = |a_d| \frac{\lambda_{X_F}(B(d, \epsilon))}{\lambda_{\mathbb{R}^d}(\phi_d(B(d, \epsilon)))} = |a_d|,$$

and by induction using (2.20),

$$\frac{\lambda_{X_F}(B(N, \epsilon))}{\lambda_{\mathbb{R}^d}(\phi_d(B(N, \epsilon)))} = |a_d|^{N-d} \tag{2.21}$$

for $N \geq d$. Also, regarding A as a linear map on $\mathbb{C}^d$,

$$\lambda_{\mathbb{R}^d}(\phi_d(B(N, \epsilon)))^2 = \lambda_{\mathbb{C}^d}\left(\{\mathbf{v} \in \mathbb{C}^d : \|A^k \mathbf{v}\|_\infty < \epsilon \text{ for } k = 0, \ldots, N - d - 1\}\right). \tag{2.22}$$

Then, using Lemma 2.11,

$$-\frac{2}{N} \log \lambda_{\mathbb{R}^d}(\phi_d(B(N, \epsilon))) = -\frac{2}{N} \log \lambda_{\mathbb{R}^d}\left(\{\mathbf{t} \in \mathbb{R}^d : \|A^k \mathbf{t}\|_\infty < \epsilon \right.$$
$$\left. \text{for } k = 0, \ldots, N - s - 1\}\right)$$
$$= -\frac{1}{N} \log \lambda_{\mathbb{C}^d}\left(\{\mathbf{t} \in \mathbb{C}^d : \|A^k \mathbf{t}\|_\infty < \epsilon \right.$$
$$\left. \text{for } k = 0, \ldots, N - s - 1\}\right)$$
$$= \frac{N-s}{N} \bar{h}_{N-s}(A) \longrightarrow 2\sum_{j=1}^{d} \log^+ |\alpha_j| \text{ as } N \to \infty.$$

Together with (2.21) this means that

$$h(T_F) = \lim_{\epsilon \searrow 0} \lim_{N \to \infty} -\frac{1}{N} \log \lambda_{X_F}(B(N, \epsilon)) = \log |a_d| + \sum_{j=1}^{d} \log^+ |\alpha_j| = m(F)$$

by Definition 1.4.

The notion of expansiveness also makes sense here, and it is convenient to rephrase it in terms of an *expansive neighbourhood*: an open neighbourhood $U \subset X_F$ of the identity 0 is an expansive neighbourhood for T_F if

$$\bigcap_{n \in \mathbb{Z}} T_F^{-n}(U) = \{0\},$$

and T_F is expansive if and only if it has an expansive neighbourhood.

Theorem 2.13. *Let $F \in \mathbb{Z}[x]$ be a polynomial with $F(0) \neq 0$. Then T_F is expansive if and only if every zero α of F has $|\alpha| \neq 1$.*

Notice that by Example 2.10[4], this generalizes Lemma 2.9 for the case of matrices in companion form to their characteristic polynomial. Theorem 2.13 may be proved in many different ways: the proof here is taken from Schmidt [Sch95, Section 6].

Proof. Define

$$\epsilon = \frac{1}{10 \left(|a_0| + \ldots + |a_d| \right)}, \tag{2.23}$$

and assume that no zero of F has unit modulus. We claim first that $B = U(1, \epsilon)$ (cf. equation (2.15)) is an expansive neighbourhood for T_F. If not, then there is a point $x = (x_n) \in X_F \backslash \{0\}$ with the property that

$$x \in \bigcap_{n \in \mathbb{Z}} T_F^n(B) = \{z \in X_F : \rho(z_k, 0) < \epsilon \text{ for all } k \in \mathbb{Z}\}. \tag{2.24}$$

Let ℓ_∞ denote the Banach space of bounded real bi-sequences $(y_n)_{n \in \mathbb{Z}}$. Since $\epsilon < \frac{1}{10}$, there is a unique point $y \in \ell_\infty$ with $y_n = x_n \mod 1$ for all $n \in \mathbb{Z}$. The choice (2.23) of ϵ implies that

$$|a_0 y_k + \ldots + a_d y_{k+d}| < \epsilon \cdot \left(\sum_{j=0}^{d} |a_j| \right)$$

$$< \frac{1}{10}$$

so

$$a_0 y_k + \ldots + a_d y_{k+d} = 0 \tag{2.25}$$

for all $k \in \mathbb{Z}$.

The shift map $\sigma : \ell_\infty \to \ell_\infty$ defined by

$$\sigma(z)_k = z_{k+1} \text{ for all } k$$

is an isometry (that is, $\sup_{k \in \mathbb{Z}} |\sigma(z)_k - \sigma(z')_k| = \sup_{k \in \mathbb{Z}} |z_k - z'_k|$).
Let

$$S = \{z \in \ell_\infty : a_0 z_k + \ldots + a_d z_{k+d} = 0 \text{ for all } k \in \mathbb{Z}\};$$

by (2.25) the non-zero point y lies in S, so S is a non-trivial closed linear subspace of ℓ_∞. The space $\mathcal{B}$ of bounded linear operators $S \to S$ is a Banach algebra; let $\mathcal{A} \subset \mathcal{B}$ be the Banach subalgebra generated by the powers of σ. The Gelfand transform $* : \mathcal{A} \to C(\mathcal{M}(\mathcal{A}))$ from $\mathcal{A}$ to the Banach algebra of continuous $\mathbb{C}$-valued functions on the space of maximal ideals of $\mathcal{A}$ is a homomorphism of Banach algebras with norm less than or equal to 1 (see Appendix B). Since both σ and σ^{-1} are isometries on S,

$$|\sigma^*(\omega)| = 1 \text{ for all } \omega \in M(\mathcal{A}).$$

On the other hand, for any $z \in S$

$$a_0 z_k + \ldots + a_d z_{k+d} = \big((a_0 + a_1\sigma + \ldots + a_d\sigma^d)(z)\big)_k,$$

so

$$S = \{z \in \ell_\infty : \big(a_0 + a_1\sigma + \ldots + a_d\sigma^d\big)(z) = 0\}.$$

It follows that for any $\omega \in M(\mathcal{A})$, $\alpha = \sigma^*(\omega)$ is a complex number with modulus 1 annihilated by F. This contradicts the assumption on F and shows that B is an expansive neighbourhood for F.

Conversely, assume that α is a zero of F with $|\alpha| = 1$ and consider again the space ℓ_∞ of bounded real sequences. Reduction mod 1 sends points in ℓ_∞ onto points in $\mathbb{T}^{\mathbb{Z}}$. Define a (potentially trivial) closed subspace $T \subset \ell_\infty$ by

$$T = \{z \in \ell_\infty : a_0 z_k + \ldots + a_d z_{k+d} = 0 \text{ for all } k \in \mathbb{Z}\}.$$

A point $z = (z_k)$ in ℓ_∞ lies in T if and only if $\mathbf{x}_{k+1}^t = A\mathbf{x}_k^t$ for all $k \in \mathbb{Z}$, where (cf. proof of Theorem 2.12) $\mathbf{x}_k = (z_k, \ldots, z_{k+d-1})$ and

$$A = \begin{bmatrix} 0 & 1 & 0 & \cdots & 0 \\ 0 & 0 & 1 & \cdots & 0 \\ \vdots & \vdots & \vdots & \ddots & \vdots \\ 0 & 0 & 0 & \cdots & 1 \\ -a_0' & -a_1' & -a_2' & \cdots & -a_{d-1}' \end{bmatrix}.$$

is the companion matrix to the monic polynomial F/a_d.

If $\alpha = \pm 1$ then A has a real eigenspace on which it acts as an isometry, so we may find a non-zero point $z \in T$ with $\|z\|_\infty < \epsilon$ for any prescribed $\epsilon > 0$. Reducing mod 1 produces a non-zero point in X_F which remains within ϵ of 0 in every coordinate. It follows that T_F cannot be expansive.

If α is not real, then let E_α and $E_{\bar{\alpha}}$ be the eigenspaces of the linear map $\mathbb{C}^d \to \mathbb{C}^d$ induced by A corresponding to the eigenvalues α and $\bar{\alpha}$ respectively. Let $E = E_\alpha \oplus E_{\bar{\alpha}} \cap \mathbb{R}^d$; in E there is an A-invariant plane on which A acts as a rotation. It follows that we may find a non-zero point $z \in T$ with $\|z\|_\infty < \epsilon$ for any prescribed $\epsilon > 0$, which shows that T_F cannot be expansive as before.

Exercise 2.6. Give a different proof of one direction in Theorem 2.13 in the following way. Use the condition that F does have a zero of unit modulus to construct a point $x = (x_k)$ in X_F that has x_k close to zero for all k, which implies that T_F is not expansive.

2.4 Periodic Points in the Dynamical Interpretation

Recall from Lemma 2.3 that if B is a non-singular integer matrix with no unit-root eigenvalues, then the number of points on $\mathbb{T}^d$ fixed by the toral endomorphism T_B^n is given by

$$|\mathrm{Per}_n(T_B)| = |\det(B^n - I)|.$$

The same argument shows that if C is any integer matrix, the map T_C of the torus has $|\det(C)|$ points in its kernel (if this quantity is non-zero).

We next generalize this to the solenoid case.

Lemma 2.14. *Let $F(x) = a_d x^d + \ldots + a_0 \in \mathbb{Z}[x]$ be a polynomial, none of whose zeros is a root of unity, with $a_0 a_d \neq 0$. The number of points with period n in the associated dynamical system is given by*

$$|\mathrm{Per}_n(T_F)| = \left| a_d^n \prod_{i=1}^{d} (\alpha_i^n - 1) \right|$$

where $\alpha_1, \ldots, \alpha_d$ are the zeros of F.

Example 2.15. [1] If $F(x) = 1 - 2x$ and $G = 2x - 1$, then it is clear from (2.14) that the action of T_F on X_F is algebraically isomorphic to the action of T_G^{-1} on X_G. They should therefore have the same number of periodic points for each period. Using Lemma 2.14 we see that

$$|\mathrm{Per}_n(T_F)| = 2^n - 1,$$

and on the other hand

$$|\mathrm{Per}_n(T_G)| = 2^n \left| \left(\tfrac{1}{2} \right)^n - 1 \right| = 2^n - 1.$$

[2] Let $F(x) = a$, a constant. Then $|\mathrm{Per}_n(T_A)| = |a|^n$ (the empty product is one).

Proof (of Lemma 2.14). Define a matrix

$$B = \begin{bmatrix} 0 & a_d & 0 & \cdots & 0 \\ 0 & 0 & a_d & \cdots & 0 \\ \vdots & \vdots & \vdots & \ddots & \vdots \\ 0 & 0 & 0 & \cdots & a_d \\ -a_0 & -a_1 & -a_2 & \cdots & -a_{d-1} \end{bmatrix}$$

(this is simply the companion matrix of F/a_d used in the proof of Theorem 2.12, multiplied by a_d). Now the points of period n under T_F are given by the solutions of the family of equations

$$a_d z_{k+d} = -a_{d-1} z_{k+d-1} + \ldots + a_0 z_k$$
$$z_{k+n} = z_k$$

for all $k \in \mathbb{Z}$. These may be written using the matrix B as

$$a_d^n \begin{bmatrix} z_k \\ \vdots \\ z_{k+d} \end{bmatrix} = B^n \begin{bmatrix} z_k \\ \vdots \\ z_{k+d} \end{bmatrix}.$$

It follows that there is a one-to-one correspondence between points of period n under T_F and points in the torus $\mathbb{T}^d$ that lie in the kernel of the map induced by the integer matrix $B^n - a_d^n I$. Writing $\mu_1 = a_d \alpha_1, \ldots, \mu_d = a_d \alpha_d$ for the eigenvalues of B, we deduce that

$$|\mathrm{Per}_n(T_F)| = \left| \prod_{i=1}^d (\mu_i^n - a_d^n) \right| = \left| a_d^n \prod_{i=1}^d (\alpha_i^n - 1) \right|,$$

which is non-zero by the assumption on the zeros of F.

Theorem 2.16. *Let $F \in \mathbb{Z}[x]$, and assume that the associated dynamical system is ergodic (so no zero of F is a root of unity).*
[1] The growth rate of periodic points exists and coincides with the topological entropy,

$$\lim_{n \to \infty} \frac{1}{n} \log |\mathrm{Per}_n(T_F)| = h(T_F).$$

[2] Lehmer's growth measure

$$\lim_{n \to \infty} \frac{|\mathrm{Per}_{n+1}(T_F)|}{|\mathrm{Per}_n(T_F)|}$$

exists if and only if T_F is expansive.

Proof. Using Lemma 2.14 and Exercise 2.5, [1] follows at once from Lemma 1.10.

Turning to [2], Lemma 2.14, Theorem 2.13 and Lemma 1.2 show that if T_F is expansive then Lehmer's growth rate measure exists. Assume therefore that F has zeros $\alpha_1, \ldots, \alpha_d$ with $|\alpha_i| = 1$ for $i = 1, \ldots, s$. Since the expression

$$\prod_{j=s+1}^d \left| \frac{\alpha_j^{n+1} - 1}{\alpha_j^n - 1} \right|$$

converges, it is enough to consider

$$\prod_{j=1}^s \left| \frac{\alpha_j^{n+1} - 1}{\alpha_j^n - 1} \right|. \tag{2.26}$$

Write $\alpha_j = e^{2\pi i \theta_j}$ for $j = 1, \ldots, s$, so each $\theta_j \in (0, 1)$ is irrational. By Dirichlet's theorem (see W. Schmidt [Sch80, Chapter II, Section 1]), there is a sequence $n_k \to \infty$ with the property that

$$[n_k \theta_j] \ll \frac{1}{n_k^{1/s}}$$

for $j = 1, \ldots, s$ where $[x]$ denotes the distance from $x \in \mathbb{R}$ to $\mathbb{Z}$. Then, for any $j = 1, \ldots, s$

$$\left| \alpha_j^{n_k} - 1 \right| = \left| e^{2\pi i n_k \theta_j} - 1 \right| \ll \frac{1}{n_k^{1/s}},$$

$$\left| \alpha_j^{n_k+1} - 1 \right| = \left| \alpha_j^{n_k} - 1 + 1 - \alpha_j^{-1} \right| \gg \left| \frac{1}{n_k^{1/s}} - |\alpha_j - 1| \right| \gg |\alpha_j - 1| \gg 1,$$

and

$$\left| \alpha_j^{n_k-1} - 1 \right| = \left| \alpha_j^{n_k} - 1 + 1 - \alpha_j \right| \gg \left| \frac{1}{n_k^{1/s}} - |\alpha_j - 1| \right| \gg |\alpha_j - 1| \gg 1.$$

It follows that

$$\left| \frac{\alpha_j^{n_k+1} - 1}{\alpha_j^{n_k} - 1} \right| \gg n_k^{1/s}$$

and

$$\left| \frac{\alpha_j^{n_k} - 1}{\alpha_j^{n_k-1} - 1} \right| \ll \frac{1}{n_k^{1/s}}.$$

Thus the expression (2.26) is infinitely often arbitrarily large and infinitely often arbitrarily small.

Exercise 2.7. The *dynamical zeta function* of a continuous map $T : X \to X$ with finitely many points of each period is defined to be

$$\zeta_T(z) = \exp \sum_{n=1}^{\infty} |\mathrm{Per}_n(T)| \times \frac{z^n}{n}.$$

This function was introduced in this setting by Artin and Mazur [AM65]; overviews of the uses to which it has been put in dynamics are in Smale's survey [Sma67, Section I.4] and (more recently) in Katok and Hasselblatt [KH95, Chapter 3, Section 1].
[1] Compute the zeta function for the map T_F when $F(x) = a$ (a constant).
[2] Compute the zeta function for the map T_F when $F(x) = x^2 - 3x + 1$.
[3] Find the smallest real pole of the zeta function of T_F, where F has no unit root zeros.

Theorem 2.16 shows that 'good' periodic point behaviour is closely connected to expansiveness for group automorphisms. A natural arithmetical family of non-expansive compact group automorphisms are the S-integer systems, and some of the results about growth rates of periodic points have been extended to that setting (see Chothi *et al.* [CEW97], [War97] and [War98a]).

3. Mahler's Measure in Many Variables

Mahler's lemma (Lemma 1.8) suggests how to generalize the measure of polynomials in one variable to several variables. In this chapter we extend some of the results of Chapter 1 to this more general measure.

Definition 3.1. The *Mahler measure* of a non-zero polynomial

$$F \in \mathbb{C}[x_1, \ldots, x_n]$$

is

$$m(F) = \int_0^1 \cdots \int_0^1 \log |F(e^{2\pi i \theta_1}, \ldots, e^{2\pi i \theta_n})| d\theta_1 \ldots d\theta_n.$$

Remark 3.2. [1] The integral is potentially singular: we return to this problem below. For now, we simply state that $m(F)$ exists for all non-zero $F \in \mathbb{C}[x_1, \ldots, x_n]$ and is non-negative for all non-zero $F \in \mathbb{Z}[x_1, \ldots, x_n]$. As usual we extend the definition to include $m(0) = \infty$.
[2] The definition is independent of any absent variables (since they integrate to 1), so it really makes sense to think of the definition as applying to the set

$$\bigcup_{n \geq 1} \mathbb{C}[x_1, \ldots, x_n]$$

of all polynomials in commuting variables.
[3] Other attempts have been made to generalize Lehmer's original measure (1.3), but these do not seem to have the right arithmetical and other properties. See Amoroso [Amo96] and Myerson [Mye84] for some work in different directions.
[4] Clearly the definition of m extends to Laurent polynomials

$$F \in \mathbb{C}[x_1^{\pm 1}, \ldots, x_n^{\pm 1}]$$

either by evaluating the integral on the Laurent polynomial or by noting that the measure of any monomial must be zero and $m(FG) = m(F) + m(G)$.

Exercise 3.1. Let

$$F(\mathbf{x}) = \sum c_{\mathbf{m}} \mathbf{x}^{\mathbf{m}} \in \mathbb{C}[x_1, \ldots, x_n],$$

where $\mathbf{x^m} = x_1^{m_1} \ldots x_n^{m_n}$. Let A be an $n \times n$ integer matrix with non-zero determinant, and define $F^{(A)}(\mathbf{x}) = \sum c_{\mathbf{m}} \mathbf{x^{mA}}$. Prove that

$$m(F) = m(F^{(A)}).$$

Exercise 3.2. Let d_i be the degree in the variable x_i for $F \in \mathbb{C}[x_1, \ldots, x_n]$. Prove a version of Mahler's inequality (cf. Exercise 1.6[2]) for many variables:

$$m(\partial F/\partial x_i) \leq m(F) + \log d_i.$$

Exercise 3.3. Generalize Exercises 1.3 and 1.11 to several variables as follows. Let

$$F(x_1, \ldots, x_n) = \sum_{h_1=0}^{d_1} \cdots \sum_{h_n=0}^{d_n} a_{h_1, \ldots, h_n} x_1^{h_1} \ldots x_n^{h_n} \in \mathbb{C}[x_1, \ldots, x_n],$$

a polynomial with degree d_i in the variable x_i for each i. The *length* of F, $L(F)$, is the sum of the absolute value of the coefficients, and the *height* of F, $H(F)$, is the maximum of the absolute value of the coefficients. Prove the following inequalities and show that equality may occur:

$$M(F) \leq L(F) \leq 2^{d_1 + \cdots + d_n} M(F), \tag{3.1}$$

and

$$\{(d_1 + 1) \ldots (d_n + 1)\}^{-1/2} M(F) \leq H(F) \leq 2^{d_1 + \cdots + d_n - \nu(F)} M(F), \tag{3.2}$$

where $\nu(F)$ is the number of variables $x_1, \ldots, x_n$ that occur in F at least to the first degree.

3.1 Explicit Values

The evaluation of higher-dimensional Mahler measures is not straightforward unless there is a dominant coefficient in a convenient position in the support of the polynomial. To illustrate some of the values that arise, consider the following examples taken from Smyth's paper [Smy81b].

Example 3.3. [1] Let $F(x_1, x_2) = 2 + x_1 + x_2$. Then, using the Jensen formula twice, we see that

$$
\begin{aligned}
m(F) &= \int_0^1 \left(\int_0^1 \log |e^{2\pi i \theta_1} + e^{2\pi i \theta_2} + 2| d\theta_1 \right) d\theta_2 \\
&= \int_0^1 \log \max\{1, |e^{2\pi i \theta_2} + 2|\} d\theta_2 \\
&= \int_0^1 \log |e^{2\pi i \theta_2} + 2| d\theta_2 = \log 2.
\end{aligned}
$$

[2] $m(1+x_1+x_2) = \dfrac{3\sqrt{3}}{4\pi} \sum\limits_{n=1}^{\infty} \dfrac{\chi(n)}{n^2}$, where $\chi(n) = \left(\dfrac{n}{3}\right)$ is the Legendre symbol

$$\chi(n) = \begin{cases} 1 & n \equiv 1 \mod 3, \\ -1 & n \equiv 2 \mod 3, \\ 0 & n \equiv 0 \mod 3. \end{cases}$$

[3] $m(1 + x_1 + x_2 + x_3) = \dfrac{7}{2\pi^2} \sum\limits_{n=1}^{\infty} \dfrac{1}{n^3}$.

[4] As $n \to \infty$,

$$m(1 + x_1 + x_2 + \ldots + x_n) = \tfrac{1}{2}\log n - \tfrac{\gamma}{2} + O\left(\tfrac{1}{\sqrt{n}}\right),$$

where γ is the Euler–Mascheroni constant. A bound of the form $\frac{1}{2}\log n + O(1)$ follows fairly easily from the central limit theorem. This more refined estimate uses a quantitative form of the central limit theorem; the evaluation of the constant term appears in the Corrigendum to Smyth [Smy81b].

[5] As $n \to \infty$,

$$m(x_0 + (1 + x_1)\ldots(1 + x_n)) = \sqrt{\dfrac{\pi n}{24}} + O(\log n).$$

See Smyth [Smy81b] for a proof (sketched in Lemma 3.4 below), and many more interesting examples.

The next exercise prepares the way for a sketch proof of Example 3.3[5], included to show how Smyth used ideas from probability to arrive at various asymptotic formulæ. The proof uses ideas about random variables that are not needed anywhere else.

Exercise 3.4. [1] Show that

$$m\left(x_0 + \prod_{j=1}^{n}(1 + x_j)\right) = \int_0^1 \ldots \int_0^1 \log^+ \prod_{j=1}^{n} \left|1 + e^{2\pi i \theta_j}\right| d\theta_1 \ldots d\theta_n.$$

[2] Prove that $|1 + e^{2\pi i\theta}| = |2\cos \pi\theta|$.
[3] Prove that

$$\int_0^1 \log |2\cos \pi\theta| d\theta = 0$$

and

$$\int_0^1 (\log |2\cos \pi\theta|)^2 d\theta = \dfrac{\pi^2}{12}.$$

The second is not easy; it may help to use the standard integral from Gradshteyn and Ryzhik [GR94, 4.224 no. 8]

$$\int_0^{\pi/2} (\log \cos x)^2 dx = \dfrac{\pi}{2}\left((\log 2)^2 + \dfrac{\pi^2}{12}\right).$$

Lemma 3.4. *As* $n \to \infty$,

$$m(x_0 + (1 + x_1) \ldots (1 + x_n)) = \sqrt{\frac{\pi n}{24}} + O(\log n).$$

Proof. Using Exercise 3.4[1] and [2], we have

$$m\left(x_0 + \prod_{j=1}^{n}(1 + x_j)\right) = \int_0^1 \ldots \int_0^1 \max\left\{0, \sum_{j=1}^{n} \log |2 \cos \pi \theta_j|\right\} d\theta_1 \ldots d\theta_n.$$

$$(3.3)$$

Let $Y_j = \frac{\sqrt{12}}{\pi} \log |2 \cos \pi \theta_j|$. By Exercise 3.4[3], $Y_1, \ldots, Y_n$ are a family of independent identically distributed random variables each with mean 0 and variance 1. Let $g(x) = \max\{0, x\}$ and write P_n for the distribution function of the random variable $\frac{1}{\sqrt{n}} \sum_{j=1}^{n} Y_j$. Then by (3.3),

$$m\left(x_0 + \prod_{j=1}^{n}(1 + x_j)\right) = \frac{\pi \sqrt{n}}{\sqrt{12}} \int_0^1 \ldots \int_0^1 \max\left\{0, \frac{1}{\sqrt{n}} \sum_{j=1}^{n} Y_j\right\} d\theta_1 \ldots d\theta_n$$

$$= \pi \sqrt{\frac{n}{12}} \int_{-\infty}^{\infty} g(x) dP_n(x).$$

Let $N(x) = \frac{1}{\sqrt{2\pi}} e^{-x^2/2}$, the standard normal distribution. Apply a form of the central limit theorem to P_n to deduce that

$$\left| \int_{-\infty}^{\infty} g(x) d\left(P_n(x) - N(x)\right) \right| = O\left(\frac{\log n}{\sqrt{n}}\right) \qquad (3.4)$$

(strictly, one needs to check that the Y_j satisfy a higher moment condition; the details are in Smyth [Smy81b]). On the other hand,

$$\int_{-\infty}^{\infty} g(x) dN(x) = \frac{1}{\sqrt{2\pi}} \int_{-\infty}^{\infty} \max\{0, x\} e^{-x^2/2} dx = \frac{1}{\sqrt{2\pi}}.$$

It follows that

$$m\left(x_0 + \prod_{j=1}^{n}(1 + x_j)\right) = \sqrt{\frac{\pi n}{24}} + O(\log n).$$

Boyd's work on numerical evidence for the connection between Mahler measures and L-functions [Boy98] has led to many interesting explicit formulæ.

Example 3.5. [1] $m(x_1 x_2 + x_1^2 x_2 + x_2 + x_1 x_2^2 + x_1) = 15 \sum_{n=1}^{\infty} \frac{a_n}{n^2}$, where the coefficients (a_n) are determined by the property

$$\sum_{n=1}^{\infty} a_n x^n = x \prod_{r=1}^{\infty} (1 - x^r)(1 - x^{3r})(1 - x^{5r})(1 - x^{15r}).$$

The symmetry of the polynomial is better revealed using Laurent polynomials: Definition 3.1 applies equally well to these (cf. Remark 3.2[4]), and then the formula becomes

$$m(1 + x_1 + x_1^{-1} + x_2 + x_2^{-1}) = 15 \sum_{n=1}^{\infty} \frac{a_n}{n^2},$$

with the coefficients determined as before. Notice that the integral reduces to the harmless looking real integral

$$m(1 + x_1 + x_1^{-1} + x_2 + x_2^{-1}) = \int_0^1 \int_0^1 \log |1 + 2 \cos 2\pi s + 2 \cos 2\pi t| ds dt.$$

[2] Boyd's work has also exhibited some unexpected identities between Mahler measures, for example

$$m(x_1 + x_1^{-1} + x_2 + x_2^{-1} + 5) = 6m(x_1 + x_1^{-1} + x_2 + x_2^{-1} + 1).$$

Exercise 3.5. If $F \in \mathbb{C}[x_1, x_2]$ and F has constant coefficient a that in absolute value exceeds the sum of the absolute values of all the other coefficients, then $m(F) = \log |a|$ (see Exercise 3.8 for a generalization).

We shall prove Example 3.3[2] along the lines of Boyd's proof in [Boy81]. Recall that we write

$$\log^+ |\lambda| = \log \max\{1, |\lambda|\}$$

and $\Re(\cdot)$, $\Im(\cdot)$ for the real and imaginary parts of a complex number.

Lemma 3.6. $m(1 + x_1 + x_2) = \dfrac{3\sqrt{3}}{4\pi} \displaystyle\sum_{n=1}^{\infty} \dfrac{\chi(n)}{n^2}.$

Proof. By Jensen's formula and a substitution,

$$J = \int_0^1 \left(\int_0^1 \log |e^{2\pi i\theta_1} + e^{2\pi i\theta_2} + 1| d\theta_2 \right) d\theta_1 = \int_0^1 \log^+ |e^{2\pi i\theta_1} + 1| d\theta_1$$

$$= \int_{-1/2}^{1/2} \log^+ |1 + e^{2\pi i\theta}| d\theta.$$

Now a simple calculation shows that $|1 + e^{2\pi i\theta}| \geq 1$ for $\theta \in [-1/2, 1/2]$ only for $-1/3 \leq \theta \leq 1/3$ (see Figure 3.1).
Hence

$$J = \int_{-1/3}^{1/3} \log |1 + e^{2\pi i\theta}| d\theta = 2 \int_0^{1/3} \log |1 + e^{2\pi i\theta}| d\theta$$

$$= 2\Re \int_0^{1/3} \log \left(1 + e^{2\pi i\theta}\right) d\theta.$$

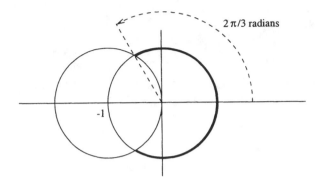

Fig. 3.1. $|1 + e^{2\pi i\theta}| \geq 1$

Now expand using a Taylor series

$$\log\left(1 + e^{2\pi i\theta}\right) = \sum_{n=1}^{\infty} (-1)^{n-1} \frac{e^{2\pi i n\theta}}{n}. \tag{3.5}$$

The series (3.5) does not converge absolutely but for $\theta \in [0, \frac{1}{3}]$ it does converge uniformly (because we are far from the singularity at $\theta = 1/2$). This may be compared with the difficult step in the proof of Jensen's formula: in evaluating $\int_0^1 \log|1 - e^{2\pi i\theta}| d\theta$ we do not have a uniformly convergent Taylor series near $\theta = 0$.

It follows that

$$J = 2\Re \sum_{n=1}^{\infty} \frac{(-1)^{n-1}}{n} \int_0^{1/3} (\cos 2\pi n\theta + i \sin 2\pi n\theta) \, d\theta$$

$$= 2 \sum_{n=1}^{\infty} \frac{(-1)^{n-1}}{n} \int_0^{1/3} \cos 2\pi n\theta \, d\theta$$

$$= 2 \sum_{n=1}^{\infty} \frac{(-1)^{n-1}}{n} \left[\frac{1}{2\pi n} \sin \frac{2\pi n}{3} \right]_0^{1/3}.$$

Notice that $\sin \frac{2\pi n}{3} = \left(\frac{n}{3}\right) \frac{\sqrt{3}}{2}$, so this reduces to

$$J = \frac{\sqrt{3}}{2\pi} \sum_{n=1}^{\infty} \frac{(-1)^{n-1}}{n^2} \chi(n)$$

$$= \frac{\sqrt{3}}{2\pi} \left\{ \sum_{n=1}^{\infty} \frac{\chi(2n-1)}{(2n-1)^2} - \sum_{n=1}^{\infty} \frac{\chi(2n)}{(2n)^2} \right\}$$

$$= \frac{\sqrt{3}}{2\pi} \left\{ \sum_{n=1}^{\infty} \frac{\chi(n)}{n^2} - 2 \sum_{n=1}^{\infty} \frac{\chi(2n)}{(2n)^2} \right\}$$

$$= \frac{\sqrt{3}}{2\pi} \left\{ \sum_{n=1}^{\infty} \frac{\chi(n)}{n^2} - \frac{1}{2}\chi(2) \sum_{n=1}^{\infty} \frac{\chi(n)}{n^2} \right\}$$

$$= \frac{\sqrt{3}}{2\pi} \frac{3}{2} \sum_{n=1}^{\infty} \frac{\chi(n)}{n^2} = \frac{3\sqrt{3}}{4\pi} \sum_{n=1}^{\infty} \frac{\chi(n)}{n^2}.$$

Series of the kind seen above are examples of L-series. Smyth, Boyd, Rodriguez Villegas and Ray have shown that many integral Mahler measures are simple multiples of values of L-series (see [Boy98], [Ray87], [Vil98] and [Smy81b]). Deninger [Den97] has given a deep – conjectural – explanation for this.

Exercise 3.6. Prove the formula in Example 3.3[3]. Hint:

$$m(1 + x_1 + x_2 + x_3) = m(1 + x_1 + x_3(1 + x_2))$$

$$= \int_0^1 \int_0^1 \log \max\{|1 + e^{2\pi i \theta_1}|, |1 + e^{2\pi i \theta_2}|\} d\theta_1 d\theta_2.$$

3.2 Existence

We now return to the question of existence for the Mahler measure.

Lemma 3.7. *The expression $m(F)$ in Definition 3.1 always exists as an improper Riemann integral. Moreover, $m(F) \geq 0$ if F has integral coefficients.*

Proof. It is clear from the triangle inequality that $m(F)$ is bounded above by $\ell(F)$ (the logarithm of the sum of the absolute values of the coefficients of F). Assume that the lemma is proved for all polynomials in $n - 1$ variables. Write F as a polynomial in x_1 with coefficients in $\mathbb{C}[x_2, \ldots, x_n]$,

$$F(x_1, \ldots, x_n) = a_d(x_2, \ldots, x_n)x_1^d + \ldots + a_0(x_2, \ldots, x_n).$$

By choosing branches of the algebraic functions arising, factorize F as follows:

$$F(x_1, \ldots, x_n) = a_d(x_2, \ldots, x_n) \prod_{j=1}^{d} (x_1 - g_j(x_2, \ldots, x_n)) \qquad (3.6)$$

for certain algebraic functions $g_1, \ldots, g_d$. Then

$$m(F) = m(a_d(x_2, \ldots, x_n)) +$$

$$\sum_{j=1}^{d} \int_0^1 \cdots \int_0^1 \underbrace{\left(\int_0^1 \log |e^{2\pi i \theta_1} - g_j(e^{2\pi i \theta_2}, \ldots, e^{2\pi i \theta_n})| d\theta_1 \right)}_{J} d\theta_2 \ldots d\theta_n$$

By the inductive hypothesis, the first term exists. Applying Jensen's formula to the indicated integral, we find that

$$J = \log^+ |g_j(e^{2\pi i \theta_2}, \ldots, e^{2\pi i \theta_n})|.$$

For each $N \in \mathbb{N}$, define

$$\alpha_N = m(a_d(x_2, \ldots, x_n))$$
$$+ \sum_{j=1}^{d} \int \cdots \int_{|g_j| \leq N} \log^+ |g_j(e^{2\pi i \theta_2}, \ldots, e^{2\pi i \theta_n})| d\theta_2 \ldots d\theta_n.$$

Then for each N, α_N exists since the integrand is continuous. Also α_N is increasing in N, and is bounded above, so

$$\alpha_N \to m(F)$$

as $N \to \infty$. If F has integral coefficients then $\alpha_N \geq 0$ (since the integrand is non-negative and $m(a_d) \geq 0$ by hypothesis).

An alternative approach to Lemma 3.7 is to analyse the nature of the (potential) singularity itself. The following result is a first step in this direction.

Write μ_n for n-dimensional Lebesgue measure on the additive torus $[0,1)^n$, normalized so that the whole torus has measure 1.

Lemma 3.8. *For small $\epsilon > 0$,*

$$\mu_n \left(\{ \boldsymbol{\theta} : |F(e^{2\pi i \theta_1}, \ldots, e^{2\pi i \theta_n})| < \epsilon \} \right) \leq C \cdot \epsilon^\delta,$$

where C and $\delta > 0$ depend only on F.

Remark 3.9. It will be clear from the proof we present that δ depends on the degree of F in each of its variables. A remarkable stronger estimate exists, which is independent of the degrees – see Schmidt [Sch95, Section 16] for the one-variable case. A proof – due to Lawton – of the stronger estimate will be given later (see Theorem 3.23 and Section 3.5 below).

Proof. Using the factorization (3.6) we see that

$$|F(\mathbf{x})| = |a_d(x_2, \ldots, x_n)| \times \prod_{j=1}^{d} |x_1 - g_j(x_2, \ldots, x_n)|,$$

so that if $|F(\mathbf{x})| < \epsilon$ then either

$$|a_d(x_2, \ldots, x_n)| < \epsilon^{1/d+1}$$

or there exists a j for which

$$|x - g_j| < \epsilon^{1/d+1}.$$

Assume the result holds for polynomials in $n - 1$ variables, so that

$$\mu_{n-1}(\{(\theta_2, \ldots, \theta_n) : |a_d(e^{2\pi i \theta_2}, \ldots, e^{2\pi i \theta_n})| < \epsilon^{1/d+1}\}) \leq C \cdot \epsilon^{\delta_1},$$

for some $\delta_1 > 0$. On the other hand,

$$\mu_1(\{\theta_1 \in [0, 1) : |e^{2\pi i \theta_1} - g_j| < \epsilon^{1/d+1}\}) \tag{3.7}$$

can only be small if $|g_j|$ is close to 1, say $g_j = re^{2\pi i \beta}$ for some r close to 1. Now

$$|e^{2\pi i \theta} - re^{2\pi i \beta}|$$

can only be small if $|2\pi i(\theta_1 - \beta)|$ is small. So the measure (3.7) depends on $\epsilon^{1/d+1}$ but not on β, and so

$$\mu_1(\{\theta_1 \in [0, 1) : |e^{2\pi i \theta_1} - g_j| < \epsilon^{1/d+1}\}) \leq c' \epsilon^{1/d+1}.$$

It follows that the worst that can happen is

$$\mu_n\left(\{\theta : |F(e^{2\pi i \theta_1}, \ldots, e^{2\pi i \theta_n})| < \epsilon\}\right) \leq \max\{c_1 \epsilon^{\delta_1}, c_2 \epsilon^{1/d+1}\},$$

which proves the lemma.

This gives an alternative approach to proving Lemma 3.7. Given $\epsilon > 0$, the integral

$$\int_{|F| \geq \epsilon}$$

clearly exists because the integrand is continuous. To show that $\int_{|F| < \epsilon}$ contributes nothing in the limit as $\epsilon \to 0$, write $\nu(x) = \mu(\{z : |F(z)| < x\})$ and integrate by parts:

$$\int_{|F| < \epsilon} = \int_0^\epsilon \log x \cdot d\nu$$

$$= \nu(x) \log x |_0^\epsilon - \int_0^\epsilon \frac{\nu(x)}{x} dx.$$

On the other hand, $\nu(x) = O(x^\delta)$ so both terms are $O(\epsilon^\delta | \log \epsilon|)$, which converges to zero as $\epsilon \to 0$.

3.3 When Does the Measure Vanish?

Just as in the single-variable case, it turns out that the integer polynomials whose Mahler measure vanishes can be described completely. First notice that Exercise 1.6[1] has a higher-dimensional analogue: if $F \in \mathbb{Z}[x]$ has $m(F) = 0$, then for any $(a, b) \neq (0, 0)$, the polynomial $F(x_1^a x_2^b) \in \mathbb{Z}[x_1, x_2]$ also has $m(F) = 0$. For example, in $\mathbb{Z}[x_1, x_2]$ any monomial has measure zero, $\phi(x_1^a x_2^b)$ has measure zero for ϕ any cyclotomic polynomial, and any product of polynomials with measure zero has measure zero. The basic result on vanishing of Mahler measures is that these examples cover essentially all the cases once we pass to Laurent polynomials (simply to avoid problems with polynomials like $x_2^2(1 + (x_1/x_2)^2)$).

Theorem 3.10. *For any primitive polynomial $F \in \mathbb{Z}[x_1^{\pm 1}, \ldots, x_n^{\pm 1}]$, $m(F)$ is zero if and only if F is a monomial times a product of cyclotomic polynomials evaluated on monomials.*

We will prove the complex analogue of this first and then use it to deduce the integral form. For simplicity we also assume that $n = 2$ and indicate at the end how the proof extends to $n > 2$. The proof given is due to Smyth [Smy81a].

Definition 3.11. A non-zero polynomial $F \in \mathbb{C}[x]$ is said to be *unit-monic* if $F(x) = a_d x^d + \ldots + a_0$ has $|a_d| = |a_0| = 1$.

For unit-monic polynomials there is a complex analogue of Kronecker's lemma.

Lemma 3.12. *If $F \in \mathbb{C}[x]$ is unit-monic then $m(F) = 0$ if and only if all zeros of F lie on the unit circle.*

Proof. This is clear.

Definition 3.13. Let F be a non-zero polynomial in $\mathbb{C}[x_1^{\pm 1}, x_2^{\pm 1}]$, written

$$F(x_1, x_2) = \sum_{j \in J} a(j) x_1^{j_1} x_2^{j_2},$$

with J finite and all $a(j) \neq 0$; the set $J = J(F)$ is the *support* of the polynomial F. Define $\mathcal{C}(F)$ to be the convex hull of $J \subset \mathbb{Z}^2$, called the *Newton polygon* of F.

Example 3.14. [1] Let $F(x_1, x_2) = 1 + x_1 + x_2$; then the Newton polygon is a triangle (see Figure 3.2).
[2] Let $G(x_1, x_2) = 1 + x_1 + x_1 x_2$; then the Newton polygon is again a triangle (see Figure 3.3).
[3] Both (a) $1 + x_1^2 + x_1 x_2$ and (b) $1 + 2x_1 + x_1^2 + 3x_1 x_2$ have the same Newton polygon (see Figure 3.4).
[4] The Newton polygon of $2 + x_1 + x_2 + 3x_1 x_2$ is a square (see Figure 3.5).

We now define the appropriate generalization of unit-monic for polynomials in two variables. Recall that an *extreme point* of the convex hull of a finite set $S \subset \mathbb{Z}^2$ is a point lying on the intersection of two faces.

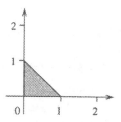

Fig. 3.2. Newton polygon for $1 + x_1 + x_2$

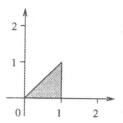

Fig. 3.3. Newton polygon for $1 + x_1 + x_1 x_2$

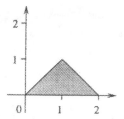

Fig. 3.4. Newton polygon for $1 + x_1^2 + x_1 x_2$ and $1 + 2x_1 + x_1^2 + 3x_1 x_2$

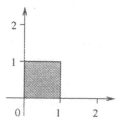

Fig. 3.5. Newton polygon for $2 + x_1 + x_2 + 3x_1 x_2$

Definition 3.15. A non-zero polynomial $F \in \mathbb{C}[x_1, x_2]$ is *extreme-monic* if $|a(\mathbf{j})| = 1$ for extreme points $\mathbf{j} \in \mathcal{C}(F)$.

For Example 3.14, [1], [2], and [3](a) are extreme-monic; the others are not.

Lemma 3.16. *If F and G are polynomials in $\mathbb{C}[x_1, x_2]$ then*

[1] $\mathcal{C}(FG) = \mathcal{C}(F) + \mathcal{C}(G)$;
[2] *every extreme point of $\mathcal{C}(FG)$ is a sum of extreme points of $\mathcal{C}(F)$ and $\mathcal{C}(G)$ in a unique way;*
[3] *if any two of F, G and FG are extreme-monic then so is the third.*

Parts [1] and [2] are geometric proofs, and [3] follows from them. To see how the proof should go, first look at the product of $F(x_1, x_2) = 1 + x_1 + x_2$ and $G(x_1, x_2) = 1 + x_1 + x_1 x_2$ from Example 3.14. Since

$$FG = 1 + 2x_1 + 2x_1 x_2 + x_1^2 + x_1^2 x_2 + x_2 + x_1 x_2^2,$$

the Newton polygon of FG is as shown in Figure 3.6.

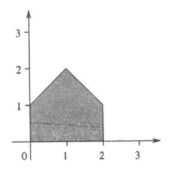

Fig. 3.6. The Newton polygon of FG

Notice that the shape of the Newton polygon can be seen directly from the two polygons without multiplying the polynomials, simply by sliding one polygon around the other, as illustrated in Figure 3.7.

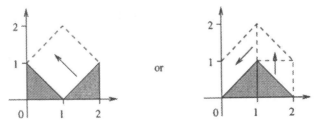

Fig. 3.7. Multiplying two polynomials

Lemma 3.17. *Suppose $F \in \mathbb{C}[x_1^{\pm 1}, x_2^{\pm 1}]$ is a non-zero polynomial. Then $\mathcal{C}(F)$ is a straight line if and only if F is a monomial times a one-variable polynomial evaluated on a monomial,*

$$F(x_1, x_2) = x_1^a x_2^b G(x_1^c x_2^d)$$

for some $G \in \mathbb{C}[x^{\pm 1}]$.

Exercise 3.7. [1] Prove Lemma 3.16.
[2] Prove Lemma 3.17.

To illustrate Lemma 3.17, consider the following examples.

Example 3.18. [1] $F(x_1, x_2) = x_1 + x_2$ has $a = 0, b = 1, c = -1, d = 1, G(x) = x + 1$.
[2] $F(x_1, x_2) = x_1 + 2x_1^2 x_2 + 7x_1^3 x_2^2 = x_2^{-1}(x_1 x_2 + 2(x_1 x_2)^2 + 7(x_1 x_2)^3)$, so $a = 0, b = -1, c = d = 1, G(x) = x + 2x^2 + 7x^3$.

It follows that we can manufacture extreme-monic Laurent polynomials in two variables with vanishing Mahler measure using the following:

- monomials;
- unit-monic polynomials in one variables with zero measure evaluated on monomials;
- products of the last two types of polynomials.

Theorem 3.19. *A polynomial $F \in \mathbb{C}[x_1^{\pm 1}, x_2^{\pm 1}]$ is extreme-monic with $m(F) = 0$ if and only if F is a monomial times a product of unit-monic measure-zero polynomials evaluated on monomials.*

In proving this theorem, we shall follow an example at various points. The difficult direction is of course to show that a polynomial with vanishing Mahler measure must be of the stated form, so the example we will follow in boxed statements is the following. The polynomial

$$E(x_1, x_2) = x_1^4 + x_1^3 x_2 + x_1^2 x_2^2 + x_2^3 + x_1 x_2^2 + x_2 x_1^2$$

has $m(E) = 0$.

Proof (of Theorem 3.19). Assume that F is a polynomial of the stated form. It is clear that $m(F) = 0$. We need to check that F is extreme-monic. By Lemma 3.17, a one-variable polynomial in a monomial has a straight line as Newton polygon. Consider now two unit-monic polynomials in monomials, F and G. If their Newton polygons are parallel then their product also has a straight-line Newton polygon so there is nothing to check (since a product of unit-monics in one variable is unit-monic). If they are not parallel, then up to translates (multiplication by monomials), $\mathcal{C}(FG)$ is obtained by translating F and G to have a common point, and then filling out the parallelogram they define. The extreme points of the Newton polygon of the product come from products of extreme points of the unit-monics, so the product is extreme-monic.

Continuing inductively, if H is a unit-monic then the Newton polygon of FGH is obtained by translating $\mathcal{C}(H)$ to have a point in common with $\mathcal{C}(FG)$ and then filling out the resulting shape.

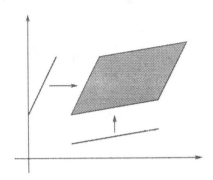

Fig. 3.8. Multiplying unit-monic polynomials

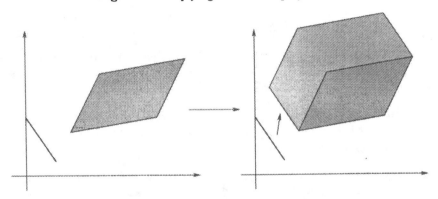

Fig. 3.9. Multiplying a unit-monic by an extreme-monic

In each case it is clear that multiplying an extreme-monic polynomial by a unit-monic one gives an extreme-monic one.

The forward direction of the theorem is more involved. Suppose F is an extreme-monic polynomial with $m(F) = 0$. Up to multiplication by a monomial,

$$F(x_1, x_2) = a_d(x_2)x_1^d + \ldots + a_1(x_2)x_1 + a_0(x_2) \in \mathbb{C}[x_2][x_1]. \qquad (3.8)$$

Lemma 3.20. *If, in* (3.8), *the polynomial* $a_d(x_2)$ *is unit-monic and not a monomial, then we may write*

$$F(x_1, x_2) = a_d(x_2)H = a_d(x_2)\left(x_1^d + b_{d-1}(x_2)x_1^{d-1} + \ldots + b_0(x_2)\right)$$

with $b_i \in \mathbb{C}[x_2]$.

By Lemma 3.16(c) we see that since F is extreme-monic, and a_d is unit-monic (hence extreme-monic), H must be extreme-monic, with $m(H) = 0$. It follows that H is an extreme-monic with $m(H) = 0$ and H *has one fewer irreducible factor than* F. Note that in $\mathbb{C}[x_1^{\pm 1}, x_2^{\pm 1}]$ monomials are units: this is important because we are going to apply transformations which preserve the number of irreducible factors of polynomials in $\mathbb{C}[x_1^{\pm 1}, x_2^{\pm 1}]$. They do not preserve the number of irreducible factors in $\mathbb{C}[x_1, x_2]$.

It follows by induction on the number of irreducible factors that H is itself a monomial times a product of unit-monics in monomials each with zero measure, and Theorem 3.19 is proved.

It remains to prove Lemma 3.20, and to show that without affecting the value of the measure or the number of irreducible factors, we may ensure that any polynomial can be put in the form (3.8) with $a_d(x)$ being unit-monic and not a monomial.

Proof (of Lemma 3.20). In equation (3.8) factor out $a_d(x_2)$ to arrive at

$$F(x_1, x_2) = a_d(x_2) \left(x_1^d + \frac{a_{d-1}(x_2)}{a_d(x_2)} x_1^{d-1} + \ldots + \frac{a_0(x_2)}{a_d(x_2)} \right) = a_d(x_2) H(x_1, x_2).$$

It follows that

$$0 = m(F) = m(a_d) + m(H), \tag{3.9}$$

with $m(a_d) \geq 0$ since $a_d(x_2)$ is unit-monic, and with $m(H) \geq 0$ by Jensen's formula:

$$H(x_1, x_2) = \prod_{j=1}^{d} (x_1 - g_j(x_2)), \tag{3.10}$$

so

$$m(H) = \sum_{j=1}^{d} \int_0^1 d \left(\int_0^1 \log |e^{2\pi i \theta_1} - g_j(e^{2\pi i \theta_2})| d\theta_1 \right) d\theta_2$$

$$= \sum_{j=1}^{d} \int_0^1 \left(\log^+ |g_j(e^{2\pi i \theta_2})| \right) d\theta_2 \geq 0.$$

Thus, (3.9) implies that

$$m(a_d) = m(H) = 0.$$

Since a_d is unit-monic, all of its roots must lie on the unit circle by Lemma 3.12.

We now claim that H is a polynomial. That is, $a_d(x_2)$ divides $a_{d-i}(x_2)$ for $i = 1, \ldots, d$. If not, then at least one of the algebraic functions g_j must have a singularity at α, a zero of $a_d(x_2)$. Write $g = g_j$. In a neighbourhood of α,

$$g(z) = \frac{c}{(z - \alpha)^q} + \ldots$$

for some rational $q \in \mathbb{Q}$ and $c \in \mathbb{C} \backslash \{0\}$. In particular, there is an open arc in $K = \mathbb{S}^1$ containing α on which $|g| > 2$. The open arc has length $\delta > 0$ say, so

$$\int_0^1 \log^+ |g(e^{2\pi i \theta_2}| d\theta_2 \geq 2\delta > 0,$$

which contradicts the fact that $m(H) = 0$. Since H is a polynomial, we have proved Lemma 3.20.

If we could be sure that $a_d(x_2)$ is always going to come out unit-monic and non-monomial then we would be done.

Example

Recall the polynomial

$$E(x_1, x_2) = x_1^4 + x_1^3 x_2 + x_1^2 x_2^2 + x_2^3 + x_1 x_2^2 + x_2 x_1^2$$

which has zero measure. In this case $a_d(x_2) = 1$ so Lemma 3.20 does not apply to E.

The proof of Theorem 3.19 is completed by showing that a harmless change of variables allows any polynomial to be put into a form in which $a_d(x_2)$ is unit-monic and non-monomial.

Choose a non-trivial (more than one point) face in the Newton polygon $\mathcal{C}(F)$. If F is a polynomial in a monomial we are done, so we may assume that the face is not all of $\mathcal{C}(F)$. Find a primitive integer vector $\mathbf{v} = \begin{bmatrix} v_{11} \\ v_{21} \end{bmatrix}$ normal to the chosen face, and complete $\mathbf{v}$ to a matrix

$$\begin{bmatrix} v_{11} & v_{12} \\ v_{21} & v_{22} \end{bmatrix} \in SL_2(\mathbb{Z}).$$

Example

The Newton polygon of E with a chosen face indicated:

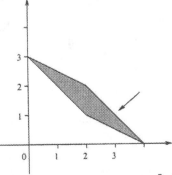

The primitive normal here may be chosen to be $\begin{bmatrix} 1 \\ 1 \end{bmatrix}$, which may be completed for example to

$$\begin{bmatrix} 1 & 0 \\ 1 & 1 \end{bmatrix} \in SL_2(\mathbb{Z}).$$

Now define a transformation from the variables x_1, x_2 into variables w_1, w_2 by

$$x_1 = w_1^{v_{11}} w_2^{v_{12}}$$
$$x_2 = w_1^{v_{21}} w_2^{v_{22}}.$$

The polynomial $G(w_1, w_2) = F(x_1, x_2)$ has the same number of irreducible factors as F did (since $G(w_1, w_2) \in \mathbb{C}[w_1^{\pm 1}, w_2^{\pm 2}]$, and we consider irreducible polynomials in that ring), and the chosen face lies on some line

$$j_1 v_{11} + j_2 v_{21} = d.$$

By convexity, all the other points of $\mathcal{C}(F)$ lie to one side of the line, so without loss of generality

$$j_1 v_{11} + j_2 v_{21} < d$$

for (j_1, j_2) in the Newton polygon of F and off the chosen face.

Example

Applying the change of variable $x_1 = w_1$, $x_2 = w_1 w_2$ from the chosen matrix

$$\begin{bmatrix} 1 & 0 \\ 1 & 1 \end{bmatrix} \in SL_2(\mathbb{Z})$$

we obtain

$$E(x_1, x_2) = w_1^4 + w_1^4 w_2 + w_1^4 w_2^2 + w_1^3 w_2^3 + w_1^3 w_2^2 + w_1^3 w_2,$$

and the point of the change of variable is that this is guaranteed to have the right form for Lemma 3.20:

$$E(x_1, x_2) = w_1^4 \left(1 + w_2 + w_2^2\right) + w_1^3 \left(w_2^3 + w_2^2 + w_2\right).$$

The result is that we should find a further factorization:

$$E(x_1, x_2) = \left(1 + w_2 + w_2^2\right) \left(w_1^4 + w_1^3 w_2\right).$$

After another step we would obtain the final form:

$$E(x_1, x_2) = \underbrace{w_2 w_1^3}_{\text{monomial}} \underbrace{\left(1 + w_2 + w_2^2\right)}_{\text{unit monic}} \underbrace{\left(1 + w_1 w_2^{-1}\right)}_{\text{unit monic in monomial}}$$

Now consider the effect of the change of variable on a typical monomial in the Newton polygon:

$$x_1^{j_1} x_2^{j_2} \longrightarrow w_1^{v_{11} j_1} w_2^{v_{12} j_1} w_1^{v_{21} j_2} w_2^{v_{22} j_2} = w_1^{v_{11} j_1 + v_{21} j_2} w_2^{v_{12} j_1 + v_{22} j_2}.$$

On the chosen face, this is

$$w_1^d w_2^{v_{12}j_1 + v_{22}j_2},$$

while *off* the leading face we must see lower powers of w_1.

So

$$G(w_1, w_2) = w_1^d a_d(w_2) + \dots$$

and the coefficients of the polynomial $a_d(w_2)$ come from the chosen face of the Newton polygon of the extreme-monic polynomial F, so $a_d(w_2)$ is unit-monic.

This completes the proof of Theorem 3.19.

Two things remain to be done: the first is to show that Theorem 3.19 implies Theorem 3.10 for $n = 2$. The second is to indicate how the method may be extended to handle more variables.

Proof (that Theorem 3.19 implies Theorem 3.10 for $n = 2$). Looking at the hypothesis for Theorem 3.10 there is no extreme-monic assumption. Given a non-zero polynomial $F \in \mathbb{Z}[x_1^{\pm 1}, x_2^{\pm 1}]$, choose a non-trivial face of the Newton polygon of F. The method of the proof shows that $m(G) = 0$ where G is the polynomial corresponding to that face.

By Lemma 3.17, G is a monomial times a polynomial in a monomial since $\mathcal{C}(G)$ is a line. The first and last coefficients of this polynomial are integers with modulus one (since $m(G) = 0$). It follows that they must be ± 1. Thus F is guaranteed to be extreme-monic since this holds for all faces of F, so we may apply Theorem 3.19.

We deduce that

$$F(x_1, x_2) = \text{monomial} \times \prod_i F_i(x_1^{n_i} x_2^{m_i})$$

for some $n_i, m_i \in \mathbb{Z}$, and the roots of each F_i have absolute value one. Note that we showed the factors F_i correspond to faces of the Newton polygon of F. Thus, they are integral polynomials. Since F is primitive, we may choose the F_i inductively to also be primitive by Gauss' lemma (see Cassels [Cas86]), which states that if F_1, F_2 are primitive, then $F_1 F_2$ is primitive also. We go on to show that they are in fact cyclotomic polynomials evaluated on the monomials $x_1^{n_i} x_2^{m_i}$. Choose $r_1, r_2 \in \mathbb{Z}$ so that $r_1 n_i + r_2 m_i \neq 0$ for all i, and let

$$x_1 = z^{r_1}, \quad x_2 = z^{r_2}$$

(this is possible since the integer vector (r_1, r_2) only has to be chosen so as to not lie on any of a finite collection of lines). Then $F(x_1, x_2)$ is a single power of z times a single-variable polynomial

$$\prod_i F_i(z^{r_1 n_i + r_2 m_i})$$

all of whose zeros have unit modulus. Apply Kronecker's lemma to deduce that all the zeros are unit roots, which proves the theorem.

The proof of Theorem 3.10 for $n > 2$ uses the same ideas but there is a double induction: when we construct the analogue of Lemma 3.20 the expression is

$$a_d(w_2, \ldots, w_n) w_1^d + \ldots$$

and so we use induction on n, the number of variables, as well as the number of irreducible factors. In addition the geometry is a little more involved: instead of choosing a linear face, we choose an $(n-1)$-dimensional face and so on. The change of variable step involves completing a primitive integral vector $\mathbf{v} \in \mathbb{Z}^n$ to an element of $SL_n(\mathbb{Z})$; this is always possible by a classical result of Hermite (see Corollary A.9 in Appendix A).

Exercise 3.8. Suppose that $F \in \mathbb{C}[x_1, x_2]$ and F has a coefficient a whose absolute value is greater than the sum of the absolute values of all the other coefficients. If this term corresponds to an extreme point of $\mathcal{C}(F)$, prove that $m(F) = \log |a|$.

For example, $m(x_1^2 + x_1 x_2 + x_1 x_2^2 + 5x_2 + 1) = \log 5$.

The next question is due to Boyd [Boy81, p. 461].

Question 5. Define the *dimension* of $F \in \bigcup_{n \geq 1} \mathbb{Z}[x_1, \ldots, x_n]$ to be $\dim(F) = \dim(\mathcal{C}(F))$, the dimension of the Newton polygon of F. Theorem 3.10 shows that an *irreducible* polynomial with $m(F) = 0$ must be one-dimensional. Define

$$\lambda(m) = \min\{m(F) : F \text{ is irreducible and } \dim(F) = m\}.$$

Is it true that $\lambda(m) \to \infty$ as $m \to \infty$?

3.4 Approximations to Two-Dimensional Mahler Measure

Lehmer's problem asked if there is a gap between 1 and $1 + \epsilon$ for some positive ϵ in the set of values of the Mahler measure of single-variable integer polynomials. The analogous question may be asked for polynomials in several variables – but this turns out to be the same question because of a surprising approximation result. We state and prove this for two-variable polynomials, following the proof of Lawton [Law83]. An alternative proof for two variables is in Boyd's paper [Boy81, Appendix 3], and there is an exposition of Lawton's proof in the general case in Schmidt's monograph [Sch95, Proposition 16.2].

Theorem 3.21. *For a non-zero polynomial $F \in \mathbb{C}[x_1, x_2]$,*

$$\lim_{N \to \infty} m(F(x, x^N)) = m(F(x_1, x_2)).$$

Before proving Theorem 3.21 we deal with a simple case and give some notation. Denote as usual the (multiplicative) circle group by $K = \mathbb{S}^1$, and the torus by $K^2 = \mathbb{S}^1 \times \mathbb{S}^1$. For an integrable function $f : K \to \mathbb{C}$, write

$$\int_0^1 f(e^{2\pi i\theta})d\theta = \int f(x)d\mu_K = \int f d\mu_K$$

and for an integrable function $g : K^2 \to \mathbb{C}$, write

$$\int_0^1 \int_0^1 g(e^{2\pi i\theta_1}, e^{2\pi i\theta_2})d\theta_1 d\theta_2 = \int g(x_1, x_2)d\mu_{K^2} = \int g d\mu_{K^2}.$$

We will use the Lebesgue measure μ_K on the circle to evaluate the measure of disjoint unions of intervals (whose measure is simply the sum of the lengths).

Lemma 3.22. Let $\phi : K^2 \to \mathbb{C}$ be any continuous function. Then

$$\lim_{N \to \infty} \int \phi(x, x^N)d\mu_K = \int \phi d\mu_{K^2}.$$

For example, if the polynomial F does not vanish anywhere on K^2, then $\log|F|$ is a continuous function so Lemma 3.22 implies Theorem 3.21 for that polynomial.

Proof. By the Stone–Weierstrass approximation theorem (see Appendix B), for any $\epsilon > 0$ there is an M and there are coefficients $\{a_{\mathbf{n}}\}_{\|\mathbf{n}\| < M}$ for which

$$\left| \phi(x_1, x_2) - \sum_{\|\mathbf{n}\| < M} a_{\mathbf{n}} x_1^{n_1} x_2^{n_2} \right| < \epsilon$$

for all $x_1, x_2 \in K$, where $\|\mathbf{n}\| = \max\{|n_1|, |n_2|\}$. Since this estimate is uniform, it is enough to show that the required convergence happens for the function

$$G(x_1, x_2) = \sum_{\|\mathbf{n}\| < M} a_{\mathbf{n}} x_1^{n_1} x_2^{n_2}.$$

Now

$$\int G d\mu_{K^2} = \sum_{\|\mathbf{n}\| < M} \int a_{\mathbf{n}} x_1^{n_1} x_2^{n_2} d\mu_{K^2} = a_0.$$

On the other hand

$$\int G(x, x^N)d\mu_K = \sum_{\|\mathbf{n}\| < M} \int a_{\mathbf{n}} \int x^{n_1 + Nn_2} d\mu_K = \sum_{\mathbf{n}:n_1 + Nn_2 = 0} a_{\mathbf{n}}.$$

For fixed M, when N is large $n_1 + Nn_2 = 0$ with $|\mathbf{n}| < M$ if and only if $n_1 = n_2 = 0$, so

$$\lim_{N \to \infty} \int G(x, x^N)d\mu_K = a_0 = \int G d\mu_{K^2}.$$

The finite sum G used to approximate ϕ is called a *trigonometric poly-nomial* since in the additive group notation the monomial $x_1^n x_2^m$ corresponds to $e^{2\pi i(nt_1 + mt_2)}$ under the correspondence $x_1 = e^{2\pi i t_1}, x_2 = e^{2\pi i t_2}$.

Exercise 3.9. Let $\phi : K^2 \to \mathbb{R}$ be any Riemann-integrable function and $\delta > 0$ be given. Show that there are finite trigonometric series $P(x_1, x_2) = \sum_{\|\mathbf{n}\| < M} a_{\mathbf{n}} x_1^{n_1} x_2^{n_2}$ and $Q(x_1, x_2) = \sum_{\|\mathbf{n}\| < M} b_{\mathbf{n}} x_1^{n_1} x_2^{n_2}$ with the property that

$$P(x_1, x_2) \leq \phi(x_1, x_2) \leq Q(x_1, x_2) \tag{3.11}$$

for all $(x_1, x_2) \in K^2$ and

$$\int (Q - P) d\mu_{K^2} < \delta. \tag{3.12}$$

Proof (of Theorem 3.21). Fix $\epsilon > 0$, and remove the potential singularities in the integrand by writing

$$\int \log |F(x, x^N)| d\mu_K = \int_{|F| < \epsilon} \log |F(x, x^N)| d\mu_K + \int_{|F| \geq \epsilon} \log |F(x, x^N)| d\mu_K. \tag{3.13}$$

Define a function $G_\epsilon : K^2 \to \mathbb{R}$ by

$$G_\epsilon(x_1, x_2) = \begin{cases} F(x_1, x_2) & \text{if } |F(x_1, x_2)| \geq \epsilon; \\ 1 & \text{if not.} \end{cases}$$

Now $\log |G_\epsilon|$ is not continuous but is Riemann-integrable (since it is continuous away from the boundary of a finite collection of open sets) so, by Exercise 3.9, given any $\delta > 0$ there are finite trigonometric series

$$P(x_1, x_2) = \sum_{|\mathbf{n}| < M} a_{\mathbf{n}} x_1^{n_1} x_2^{n_2} \text{ and } Q(x_1, x_2) = \sum_{|\mathbf{n}| < M} b_{\mathbf{n}} x_1^{n_1} x_2^{n_2}$$

with the property that

$$P(x_1, x_2) \leq \log |G_\epsilon(x_1, x_2)| \leq Q(x_1, x_2) \tag{3.14}$$

for all $(x_1, x_2) \in K^2$ and

$$\int (Q - P) d\mu_{K^2} < \delta. \tag{3.15}$$

By construction,

$$\int \log |G_\epsilon(x, x^N)| d\mu_K = \int_{|F| \geq \epsilon} \log |F(x, x^N)| d\mu_K$$

and

$$\int \log |G_\epsilon(x_1, x_2) d\mu_{K^2} = \int_{|F| \geq \epsilon} \log |F(x_1, x_2)| d\mu_{K^2}$$

so by Lemma 3.22

$$\int P(x, x^N) d\mu_K < \int_{|F| \geq \epsilon} \log |F(x, x^N)| d\mu_K < \int Q(x, x^N) d\mu_K, \qquad (3.16)$$

$$\int P(x, x^N) d\mu_K \to \int P(x_1, x_2) d\mu_{K^2}, \qquad (3.17)$$

and

$$\int Q(x, x^N) d\mu_K \to \int Q(x_1, x_2) d\mu_{K^2}. \qquad (3.18)$$

Since $\int Q d\mu_{K^2} - \int P d\mu_{K^2} < \delta$ and $\delta > 0$ was arbitrary, (3.16), (3.17) and (3.18) then show that

$$\int_{|F| \geq \epsilon} \log |F(x, x^N)| d\mu_K \longrightarrow \int_{|F| \geq \epsilon} \log |F(x_1, x_2)| d\mu_{K^2}.$$

By the estimate in Lemma 3.8,

$$\left| \int_{|F| \geq \epsilon} \log |F(x_1, x_2)| d\mu_{K^2} - \int \log |F(x_1, x_2)| d\mu_{K^2} \right| = O(\epsilon |\log \epsilon|),$$

so to complete the proof we need only show that the first term in equation (3.13) goes to zero in ϵ. The problem is that the estimate from Lemma 3.8 is of the form

$$\mu \left(\{x \in K : |F(x, x^N)| < \epsilon\} \right) < C \cdot \epsilon^{1/d}$$

where d is the degree of $F(x, x^N)$ in x, which goes to infinity with N. Thus, as $N \to \infty$ the quantity $\epsilon^{1/d}$ does not remain small. We clearly need a more sophisticated bound.

Theorem 3.23. *Suppose $F \in \mathbb{C}[x]$ is a non-zero polynomial with $k \geq 2$ non-zero terms. Then*

$$\mu \left(\{z \in K : |F(z)| < \epsilon\} \right) \leq C_k \cdot \epsilon^{1/k-1},$$

where C_k is a bounded multiple of k and the reciprocal of the leading coefficient of F.

Now for large N, the number of non-zero coefficients in $F(x, x^N)$ is constant, say k. In order to estimate the first term in (3.13), write

$$\psi(\epsilon) = \mu \left(\{z \in K : |F(z, z^N)| < \epsilon\} \right).$$

Then Theorem 3.23 says that

$$\psi(\epsilon) \leq C_k \cdot \epsilon^{1/k-1}. \tag{3.19}$$

For fixed $\epsilon \in (0,1)$, integration by parts gives

$$\int_{|F(z,z^N)|<\epsilon} \log|F(z,z^N)|d\mu_K = \int_0^\epsilon \log t\, d\psi(t)$$
$$= \psi(\epsilon)\log\epsilon - \int_0^\epsilon \psi(t)\frac{1}{t}dt$$
$$\geq C_k \cdot \epsilon^{1/k-1}\log\epsilon - \int_0^\epsilon C_k \cdot t^{(1/k-1)-1}dt$$
$$= C_k\left(\epsilon^{1/k-1}\log\epsilon - (k-1)\epsilon^{1/k-1}\right)$$
$$\to 0 \text{ as } \epsilon \to 0$$

which completes the proof of Theorem 3.21.

Exercise 3.10. Give a different proof of Theorem 3.21 using Theorem 3.23 on the following lines. It is enough to show that

$$\limsup_{N\to\infty}\left|\int \log|F(x,x^N)|d\mu_K - \int \log|F(x_1,x_2)|d\mu_{K^2}\right| = 0.$$

Prove this using a continuous function ϕ_ϵ that is equal to 1 where $|F| \geq \epsilon$ and is equal to 0 where $|F| < \epsilon/2$.

3.5 Lawton's Estimate

Before proving Lawton's estimate Theorem 3.23, consider the following examples.

Example 3.24. [1] Let $F(z) = (z-1)^d$. By the binomial theorem the number of non-zero coefficients is $k = d+1$. Now

$$|z-1|^d < \epsilon \iff |z-1| < \epsilon^{1/d},$$

so if $z = e^{2\pi i\theta}$ and $|F(z)| < \epsilon$, then $2\pi\theta$ is constrained to lie in a window of width $(2\epsilon)^{1/d}$. So

$$\mu\left(\{z \in K : |F(z)| < \epsilon\}\right) \sim \tfrac{1}{\pi}\epsilon^{1/d} = \tfrac{1}{\pi}\epsilon^{1/(k-1)}.$$

This suggest that the bound in Theorem 3.23 cannot be improved in general. [2] Let $F(z) = z^d - 1$. Now $k = 2$, and if $|z| = 1$ then $|z^d - 1| < \epsilon$ requires z to lie in one of d intervals on K, each of which covers $\frac{\epsilon}{d\pi}$ in proportion of the circle, centred at the dth roots of unity. So

$$\mu\left(\{z \in K : |F(z)| < \epsilon\}\right) \sim \frac{d}{\pi} \cdot \frac{\epsilon}{d} = \frac{\epsilon}{\pi},$$

and here $\frac{1}{k-1} = 1$.

It would simplify the proof of Theorem 3.21 if, for any polynomial $F \in \mathbb{C}[x, y]$ in which both variables appear,

$$\mu\left(\{z \in K : |F(z, z^N)| < \epsilon\}\right)$$

converged to zero in N for fixed small ϵ. The next example shows that this cannot happen.

Example 3.25. Let $F(z) = z^d + z + 1$. If $d \equiv 2 \bmod 3$ then

$$\epsilon^2 \ll \mu\left(\{z \in K : |F(z)| < \epsilon\}\right) \ll \epsilon^2.$$

To see this, argue as follows. If $d \equiv 2 \bmod 3$ then the roots of $F(z) = 0$ on the unit circle are at $\omega = e^{2\pi i/3}$ and $\bar{\omega}$. For $r = \pm 1, \pm 2, \ldots, \pm \lfloor d\epsilon \rfloor$, define

$$z_r = e^{2\pi i(1/3 + r/d)}.$$

Then for each r, $F(z_r) = O(\epsilon)$ uniformly. Moreover, around each z_r there is a neighbourhood of measure $\frac{\epsilon}{d}$ on which $F(z) = O(\epsilon)$. To see why this is so, write

$$z_r(\theta) = e^{2\pi i(1/3 + r/d + \theta)}$$

for $|\theta| \leq \frac{\epsilon}{2d}$. Then the neighbourhoods defined by letting θ vary around each z_r are disjoint, and F is still uniformly $O(\epsilon)$ on each (cf. Figure 3.10).

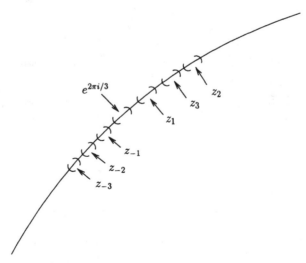

Fig. 3.10. The set on which $|F(z)| < \epsilon$

This gives approximately $d\epsilon$ intervals, each of measure roughly $\frac{\epsilon}{d}$, so there is a lower bound of the shape $c\epsilon^2$ with a uniform c.

It will be convenient to write $A(\epsilon)$ for the set of points on K^2 where $|F| < \epsilon$.

To get the upper bound, note that $|F(z)| < \epsilon$ implies $1 - \epsilon < |z+1| < 1 + \epsilon$, so $\mu(A(\epsilon)) < \epsilon$ at once. To refine this further, notice that for $z = e^{2\pi i(1/3+\theta)}$, if $|F(z)| = O(\epsilon)$ then $d\theta \bmod 1$ is $O(\epsilon)$ so $\theta - r/d = O(\epsilon/d)$ for some $r \in \mathbb{Z}$. Clearly $r = O(d\epsilon)$ giving an upper bound of $O(\epsilon^2)$.

Exercise 3.11. Let $F(z) = (z^d - 1)^{k-1}$. Show that

$$\epsilon^{1/(k-1)} \ll \mu(\{z \in K : |F(z)| < \epsilon\}) \ll \epsilon^{1/(k-1)}.$$

The full picture is not clear: the next question is entirely open.

Question 6. Let $F \in \mathbb{C}[x, y]$ be a polynomial in which both variables appear, and assume that $F(z, z^d) = 0$ has a solution on K. Is there a rational $q > 0$ (depending on F) for which

$$\frac{C_1}{k}\epsilon^q < \mu(\{z \in K : |F(z, z^d)| < \epsilon\}) < C_2 k \epsilon^q,$$

where k is the number of non-zero coefficients in $F(z, z^d)$ and C_1, C_2 are absolute constants?

There has been extensive research on the related question of the size of the set on which a polynomial can be small; see Lubinsky [Lub97] for a recent overview.

The structure of the proof of the estimate is summarized below.

Structure of proof of Theorem 3.23

(i) Define $A(\epsilon) = A(\epsilon)(F)$ and $B(\delta) = B(\delta)(F)$ and note equation (3.21):

$$\mu_\epsilon(F) = \mu(A(\epsilon)) \leq \mu(A(\epsilon) \cap B(\delta)) + \mu(B(\delta)').$$

(ii) Assume that $\mu_\epsilon(F) < E_1(k, \epsilon)$ for all polynomials F with k non-zero coefficients, where E_1 is the desired estimate.
(iii) Use (ii) to bound $\mu(B(\delta)(F')')$ in terms of $E_1(k, \epsilon)$ where F is a polynomial with $(k+1)$ non-zero coefficients and $F(0) \neq 0$.
(iv) Prove an estimate of the form

$$\mu(A(\epsilon)(F) \cap B(\delta)(F)) < E_2(k+1, \epsilon)$$

for F with $(k+1)$ non-zero coefficients.
(v) Choose δ to make the bound for $\mu(B(\delta)')$ obtained in (iii) agree with $E_2(k+1, \epsilon)$ for F with $(k+1)$ non-zero coefficients.
(vi) Deduce by induction on k that

$$\mu_\epsilon(F) < E_1(F, \epsilon)$$

for all F, with E_1 depending only on the number of non-zero coefficients in F and ϵ.

Proof (of Theorem 3.23). We may assume that F is monic, $F \in \mathbb{R}[x]$ (since $|\Re F(x)| \leq |F(x)|$), and that $F(0) \neq 0$. The last assumption is harmless since $|F(z)| = |z^{\ell} F(z)|$ for all $z \in K$, but is important because we will use induction on the number of non-zero coefficients and we need to be certain that F' has one fewer non-zero coefficient than F.

Lemma 3.26. *Given $\epsilon > 0$ and $F \in \mathbb{C}[x]$ of degree d, the set*

$$A(\epsilon) = \{ z \in K : |F(z)| < \epsilon \}$$

consists of the union of at most d intervals.

In general, for polynomials of the shape $F(z, z^d)$ there seem to be more connected components in $A(\epsilon)$ than might be expected. As we have seen in Example 3.25, $z^d + z + 1$ has at most two zeros on K, but for small ϵ there are at least $4\lfloor d\epsilon \rfloor$ connected components when $d \equiv 2$ mod 3.

Proof (of Lemma 3.26). If $|F(z)| = \epsilon$ for $z \in K$, then

$$F(z)F(\bar{z}) = F(z)F(\frac{1}{z}) = \epsilon^2,$$

so

$$z^d F(z)F(\tfrac{1}{z}) = \epsilon^2 z^d,$$

a polynomial equation of degree $2d$. It follows that there are at most $2d$ points on K where $A(\epsilon)$ and its complement meet.

Continuing with the proof of Theorem 3.23, fix $\epsilon > 0$ and define sets

$$A(\epsilon) = \{ z \in K : |F(z)| < \epsilon \},$$

$$B(\delta) = \{ z \in K : |F'(z)| \geq \delta \},$$

where $\delta > 0$ is a constant depending only on ϵ and d, the degree of F, to be chosen later.

Notice that

$$A(\epsilon) = (A(\epsilon) \cap B(\delta)) \bigsqcup (A(\epsilon) \cap B(\delta)'), \tag{3.20}$$

where $B(\delta)' = K \backslash B(\delta)$ is the complement of $B(\delta)$ and the $\sqcup$ denotes a disjoint union. It follows that

$$\mu(A(\epsilon)) = \mu(A(\epsilon) \cap B(\delta)) + \mu(A(\epsilon) \cap B(\delta)') \leq \mu(A(\epsilon) \cap B(\delta)) + \mu(B(\delta)'). \tag{3.21}$$

We want to apply induction to

$$B(\delta)' = \{ z \in K : |F'(z)| < \delta \};$$

in order to do this first notice that we may divide by z as often as needed to ensure that $F'(0) \neq 0$ without affecting the set $B(\delta)'$, and may divide by d throughout to arrive at

$$B(\delta)' = \left\{ z \in K : \left| \frac{F'(z)}{d} \right| < \frac{\delta}{d} \right\}$$

with $\frac{F'(z)}{d}$ a monic polynomial.

Now apply an inductive hypothesis: assume Theorem 3.23 for polynomials with k non-zero terms. Applying this to $\frac{F'(z)}{d}$, we have

$$\mu\left(B(\delta)'\right) < C_k \left(\frac{\delta}{d} \right)^{1/k-1}. \tag{3.22}$$

Looking at equation (3.21), we need to bound

$$\mu\left(A(\epsilon) \cap B(\delta)\right).$$

Lemma 3.27. *If $S \subset K$ is a union of m disjoint intervals, and $T \subset K$ is a union of n disjoint intervals, then $S \cap T$ is a disjoint union of no more than $m + n - 1$ intervals.*

Proof. This is clear (by induction). $\quad\blacksquare$

Now by Lemma 3.26 $A(\epsilon)$ is a union of no more than d sets, and $B(\delta)$ is a disjoint union of no more than $d - 1$ sets. It follows that $A(\epsilon) \cap B(\delta)$ is a union of no more than $2d$ disjoint intervals. Since

$$\delta \cdot \mu\left(A(\epsilon) \cap B(\delta)\right) < \int_{A(\epsilon) \cap B(\delta)} |F'| d\theta, \tag{3.23}$$

it is enough to find a suitable bound for the integral of $|F'|$.

Now

$$|F'| \leq |\Re(F')| + |\Im(F')|;$$

also $|F'(z)| = \text{sign}(F'(z))F'(z)$, and the integral of F' is easy to estimate.

Lemma 3.28. *There are no more than $2d$ intervals on which each of the functions $\Re(F')$ and $\Im(F')$ are of constant sign.*

Proof. On K, $\Re(F') = 0$ if and only if

$$F'(z) + F'(\bar{z}) = 0,$$

which is equivalent to

$$F'(z) + F'\left(\frac{1}{z}\right) = 0,$$

or (after multiplying by z^{d-1})

$$z^{d-1} F'(z) + z^{d-1} F'\left(\frac{1}{z}\right) = 0,$$

a polynomial equation of degree $2d - 2$. So there are at most $2d - 2 < 2d$ intervals of constant sign; a similar argument holds for $\Im(F)$.

It follows that there is a partition of $A(\epsilon) \cap B(\delta)$ into no more than $6d$ intervals, on each of which both $\Re(F')$ and $\Im(F')$ have constant sign. Let J denote one of these intervals. Then

$$\delta\mu(J) \leq \int_J |F'|d\theta \leq \int_J |\Re(F')|d\theta + \int_J |\Im(F')|d\theta$$
$$= \int_J \text{sign}(\Re(F')) \cdot \Re(F')d\theta + \int_J \text{sign}(\Im(F')) \cdot \Im(F')d\theta$$
$$= \text{sign}(\Re(F')) \cdot \Re(F)|_J + \text{sign}(\Im(F')) \cdot \Im(F)|_J$$
$$< 4\epsilon,$$

since both $|\Re(F)|$ and $|\Im(F)|$ are both bounded above by $|F| \leq \epsilon$ on $J \subset A(\epsilon)$. It follows that

$$\mu(J) < \frac{4\epsilon}{\delta}.$$

Since there are at most $6d$ such intervals,

$$\mu\left(A(\epsilon) \cap B(\delta)\right) \leq 24d\frac{\epsilon}{\delta},$$

so by (3.21),

$$\mu\left(A(\epsilon)\right) \leq 24d\frac{\epsilon}{\delta} + C_k \left(\frac{\delta}{d}\right)^{1/k-1}. \tag{3.24}$$

Now choose $\delta = d\epsilon^{(k-1)/k}$ to ensure that

$$\frac{\epsilon d}{\delta} = \left(\frac{\delta}{d}\right)^{1/k-1}.$$

With this choice of δ,

$$\mu\left(A(\epsilon)\right) \leq \frac{\epsilon d}{d\epsilon^{(k-1)/k}}(24 + C_k) = \epsilon^{1/k}(24 + C_k).$$

Thus the theorem is proved, with the constant C_k determined by the recurrence

$$C_{k+1} = 24 + C_k$$

(so $C_{k+1} \leq 24(k+1)$).

Exercise 3.12. [1] Let $F_d(z) = z^d + z + 1$. Prove that F_d is not cyclotomic unless $d = 2$.
[2] Let $F_d(z) = z^d - G(z)$ for fixed $G \in \mathbb{Z}[z]$. Prove that for all sufficiently large d, the zeros of F_d lie in $\{0\} \cup K$ only if $G(z) = \pm z^k$ for some $k \in \mathbb{Z}$.

The statements in the last exercise are special cases of a result of Dobrowolski, Lawton and Schinzel [DLS83]: a Laurent polynomial F in $\mathbb{Z}[y^{\pm 1}, z^{\pm 1}]$, after substituting $y = z^d$, has all zeros in $\{0\} \cup K$ for all large d if and only if F is a monomial times a product of cyclotomic Laurent polynomials evaluated on monomials. For example, $F(y, z) = y - z$ has the stated property and may be written $F(y, z) = z(yz^{-1} - 1)$.

Exercise 3.13. Use the limit formula Theorem 3.21 and the bound (1.9) from Exercise 1.4[2] to show that if $m(F(z, y)) = 0$ then $m(F(z, z^d)) = 0$ for all large d.

Exercise 3.14. Assume that $F \in [z, y]$ is monic in y. Then if $F(z, z^d)$ is cyclotomic for all large d, show that $F(z, y)$ must be a cyclotomic polynomial in y.

The next question is more involved.

Question 7. Can Exercise 3.13 be used to give an alternative proof of Theorem 3.10? All the ideas needed are in the paper of Dobrowolski, Lawton and Schnizel [DLS83].

3.6 Boyd's Example

The basic convergence result Theorem 3.21 is on the one hand a very general statement about continuous functions (by Lemma 3.22; this is really a statement about weak convergence of measures) and on the other a very specific and rather surprising statement about polynomials.

The next example is a function on the torus $\mathbb{T}^2$, continuous except at one point, for which Theorem 3.21 fails in a very strong way, showing that pushing the result further than continuous functions requires special properties of the functions considered (for example, that they be logarithms of polynomials).

Example 3.29. Define $G : \mathbb{T} \to \mathbb{C}$ by $G(\theta) = e^{2\pi i \theta \cdot 2^k}$ for $\theta \in [1 - \frac{2}{2^k}, 1 - \frac{1}{2^k})$. It is easy to check that G is continuous on $[0, 1)$ but $G(0) \neq \lim_{\theta \to 1} G(\theta)$, so G has one discontinuity on the additive circle $\mathbb{T}$.

Define $F : \mathbb{T}^2 \to \mathbb{C}$ by

$$F(\theta_1, \theta_2) = G(\theta_1) - e^{2\pi i \theta_2}.$$

Notice that for $n = 2^k$ and $\theta \in [1 - \frac{2}{2^k}, 1 - \frac{1}{2^k})$, we have $F(\theta, n\theta) = 0$. It follows that for $n = 2^k$,

$$\int_0^1 \log |F(\theta, n\theta)| d\theta = -\infty.$$

On the other hand, by Jensen's formula

$$\int_0^1 \int_0^1 \log |F(\theta_1, \theta_2)| d\theta_1 d\theta_2 = \int_0^1 \int_0^1 \log |e^{2\pi i \theta_2} - G(\theta_1)| d\theta_2 d\theta_1$$
$$= \int_0^1 \log \max\{1, |G(\theta_1)|\} d\theta_1 = 0.$$

4. Higher-Dimensional Dynamical Systems

In Chapter 2 a dynamical system was associated to each integer polynomial F in one variable. The dynamical system was given by an automorphism T_F of a compact group X_F (in the general case of Section 2.3). The map $X_F \times \mathbb{Z} \to X_F$ defined by $(x, n) \mapsto T_F^n x$ is then an *action* of $\mathbb{Z}$ by automorphisms of X_F. In this chapter we introduce higher-dimensional dynamical systems (actions of $\mathbb{Z}^n$ for $n \geq 2$) associated to polynomials in several variables.

Just as in Chapter 2, the main result is that the topological entropy of the dynamical system associated to a given polynomial is equal to the Mahler measure of the polynomial. This result was first shown by Lind, Schmidt and Ward [LSW90, Theorem 3.1] and was part of a broad development in higher-dimensional algebraic dynamical systems (see Kitchens and Schmidt [KS89], and Schmidt's monograph [Sch95]). The proof is non-trivial and only special cases will be covered here.

A $\mathbb{Z}^n$-action on a set X is a group action $T : X \times \mathbb{Z}^n \to X$; a *period* is a finite-index subgroup $\Lambda \subset \mathbb{Z}^n$, and the set of points with period Λ is

$$\mathrm{Per}_\Lambda(T) = \{x \in X : T_\mathbf{n}(x) = x \text{ for all } \mathbf{n} \in \Lambda\}.$$

4.1 The Dynamical System Associated to a Polynomial

Fix a polynomial $F \in \mathbb{Z}[x_1^{\pm 1}, \ldots, x_n^{\pm 1}]$, not identically zero. Write

$$F(\mathbf{x}) = \sum_{\mathbf{m} \in J(F)} c(\mathbf{m})\mathbf{x}^\mathbf{m}$$

where $J(F)$ is the support of F and $\mathbf{x}^\mathbf{m} = x_1^{m_1} \ldots x_n^{m_n}$. Associate to each such polynomial a compact group

$$X_F = \left\{ z = (z_\mathbf{k}) \in \mathbb{T}^{\mathbb{Z}^n} : \sum_{\mathbf{m} \in J(F)} c(\mathbf{m})z_{\mathbf{k}+\mathbf{m}} = 0 \text{ for all } \mathbf{k} \in \mathbb{Z}^n \right\}, \quad (4.1)$$

where as in (2.14) the addition is performed in $\mathbb{T}$, that is, mod 1. The group X_F carries a natural $\mathbb{Z}^n$-action T_F defined by the shift: $(z, \mathbf{n}) \mapsto (z_{\mathbf{k}+\mathbf{n}})_\mathbf{k}$. Equivalently, we may write

$$(T_{F,\mathbf{n}}(z))_{\mathbf{k}} = z_{\mathbf{k}+\mathbf{n}}.$$

Notice that for $n = 1$ this is exactly the dynamical system constructed in Section 2.3 with $T_{F,1} = T_F$. A basis of open balls around the identity in X_F is given by the sets of the form

$$B(N, \epsilon) = \{z \in X_F : \rho(z_k, 0) < \epsilon \text{ for } \|\mathbf{k}\| = 0, \ldots, N - 1\}$$

where $\|\cdot\|$ is the max norm on $\mathbb{Z}^n \subset \mathbb{R}^n$. A metric on X_F may be defined by analogy with (2.16):

$$d(z, z') = \sum_{\mathbf{k}} 2^{-\|\mathbf{k}\|} \rho(z_{\mathbf{k}}, z'_{\mathbf{k}}). \tag{4.2}$$

Example 4.1. [1] If $n = 1$ then T_F is the one-dimensional dynamical system defined in Chapter 2.

[2] If $F = a$, a constant, then

$$X_F = \prod_{\mathbb{Z}^n} \left\{0, \frac{1}{|a|}, \frac{2}{|a|}, \ldots, \frac{|a| - 1}{|a|}\right\}$$

and $T_{F,\mathbf{n}}$ is the shift by $\mathbf{n}$. If $\Lambda \subset \mathbb{Z}^n$ has index $M < \infty$ then there are $|a|^M$ points of period Λ.

[3] If $n = 2$ and $F(x_1, x_2) = x_1^2 - 5x_1 + 1$ then the condition in (4.1) is

$$z_{(k_1,k_2)} - 5z_{(k_1+1,k_2)} + z_{(k_1+2,k_2)} = 0$$

for all $(k_1, k_2) \in \mathbb{Z}^2$. For each fixed k_2, the slice through X_F is identical to the group defined in the single-variable case by $x^2 - 5x + 1$ (cf. Example 2.10[1]). So there is a group isomorphism

$$X_F \cong \left(\mathbb{T}^2\right)^{\mathbb{Z}},$$

sending the transformation $T_{F,(1,0)}$ to the infinite product of toral automorphisms $T_A^{\mathbb{Z}}$ where $A = \begin{bmatrix} 0 & 1 \\ -1 & 5 \end{bmatrix}$, and sending the transformation $T_{F,(0,1)}$ to the shift map with alphabet $\mathbb{T}^2$.

[4] If $n = 2$ and $F(x_1, x_2) = 1 + x_1 + x_2$, then

$$X_F = \{z \in \mathbb{T}^{\mathbb{Z}^2} : z_{(j,k)} + z_{(j+1,k)} + z_{(j,k+1)} = 0 \text{ for all } j, k \in \mathbb{Z}\},$$

and the action T_F is the shift action.

4.2 Mahler Measures as Entropies

The volume-growth measure of topological entropy (2.12) has a natural extension to $\mathbb{Z}^n$-actions. Let $F \in \mathbb{Z}[x_1^{\pm 1}, \ldots, x_n^{\pm 1}]$ be a non-zero polynomial, with associated group X_F, Haar measure λ and $\mathbb{Z}^n$-action T_F. The entropy is defined to be

$$h(T_F) = \lim_{\epsilon \searrow 0} \liminf_{m \to \infty} -\frac{1}{m^n} \log \lambda \left(\bigcap_{0 \le k_i < m; 1 \le i \le n} T_{F, -\mathbf{k}} \left(B(\epsilon) \right) \right) \qquad (4.3)$$

where $B(\epsilon)$ is an ϵ neighbourhood of the identity.

Theorem 4.2. *The entropy of T_F is given by*

$$h(T_F) = m(F).$$

Proof. This was first proved in Lind, Schmidt and Ward [LSW90]. A thorough exposition, with detailed discussion of important examples, is in Schmidt's monograph [Sch95].

The proof is too long to give here – and involves another (equivalent) notion of entropy defined via *separating sets*. For the simplest examples the calculations can be done directly.

Just as in the discussion after Theorem 2.6, this relates the set of possible values of $\mathbb{Z}^n$-actions by automorphisms of compact groups to Lehmer's problem (cf. Remark 1.7[3] and the discussion at the start of Section 3.4). The generalization of Lind's result [Lin77, Theorem 9.3] is shown in Lind, Schmidt and Ward [LSW90, Theorem 4.6]: the set of possible entropies of $\mathbb{Z}^n$-actions by automorphisms of compact groups is either the countable set $\{m(F) : F \in \mathbb{Z}[x_1^{\pm 1}, \ldots, x_n^{\pm 1}]\}$ or all of $[0, \infty]$ depending on the solution to Lehmer's problem. The higher-dimensional dynamical systems studied here turn out to be measure-theoretically the same as higher-dimensional Bernoulli shifts when F is irreducible (see Rudolph and Schmidt [RS95]), unless the polynomial has zero Mahler measure, so Lehmer's problem again is equivalent to asking if every higher-dimensional Bernoulli shift has an algebraic model. Lehmer's problem also appears in studying the entropy of automorphisms of 'large' dynamical systems, but there reliance on the conjecture can be avoided – see Morris and Ward [MW98].

The Mahler measure also arises in several other settings. In combinatorics, it appears as the entropy of a natural $\mathbb{Z}^n$-action on the space of essential spanning forests in a periodic graph with vertex set $\mathbb{Z}^n$. This was shown by Burton and Pemantle in [BP93] and later Solomyak gave a direct argument that the two entropies must coincide [Sol98]. Mahler's basic inequalities (cf. Remark 1.5) have been used in transcendence theory (see Philippon [Phi86] and Soulé [Sou91]). Certain exactly solved models in statistical physics have Mahler measures as physical invariants (see Kasteleyn [Kas61], Lieb [Lie67], and Temperley, Lieb and Fisher [Fis61], [TF61], [TL71]). A Mahler measure appears as the optimal constant in a sharp polynomial inequality (see Boyd [Boy92]). The (single-variable) Mahler measure is used in computer science to provide bounds on the coefficients of possible factors of polynomials (see the solution to Knuth [Knu81, Exercise 4.6.2-20]). A Mahler measure appears as the entropy of the Gaussian integer continued fraction map (see A. Schmidt [Sch93]). Mahler measures appear in the definition of the 'canonical height' (cf. Chapter 5) for hypersurfaces in toric varieties (see Maillot [Mai97]). Finally, Mahler measures arise in the study of quasi-periodic potentials for the Schrödinger equation (see Sarnak [Sar82] and Thouless [Tho90]).

An analogous construction may be made for polynomials with coefficients lying in the ring of integers of a number field – see Einsiedler *et al.* for the details [Ein], [EW97].

Example 4.3. [1] If $F(x_1, \ldots, x_n) = a$, a constant, then (cf. Example 4.1[2])

$$X_F = \prod_{\mathbb{Z}^n} \left\{ 0, \frac{1}{|a|}, \frac{2}{|a|}, \ldots, \frac{|a|-1}{|a|} \right\}$$

and $T_{F,\mathbf{n}}$ is the shift by $\mathbf{n}$. The Haar measure λ on X_F is simply the product measure. A ball of radius ϵ about the identity is a set

$$B_\epsilon = \{ z \in X_F : z_{\mathbf{k}} = 0 \text{ for } \|\mathbf{k}\| \le R(\epsilon) \}$$

for some $R(\epsilon) \to \infty$ as $\epsilon \to 0$. Then

$$\lambda(B) = \left(\frac{1}{|a|} \right)^{(2R(\epsilon)+1)^n}.$$

Now

$$\bigcap_{0 \le k_i < m; 1 \le i \le 2} T_{F,-\mathbf{k}}(B_\epsilon) = \{ z \in X_F : z_{\mathbf{k}} = 0 \text{ for } -R(\epsilon) \le k_i \le R(\epsilon) + m - 1 \}$$

so

$$\lambda \left(\bigcap_{0 \le k_i < m; 1 \le i \le n} T_{F,-\mathbf{k}}(B_\epsilon) \right) = \left(\frac{1}{|a|} \right)^{(2R(\epsilon)+m)^n}$$

giving

$$h(T_a) = \log|a| = m(a).$$

[2] If $n = 2$ and $F(x_1, x_2) = x_1^2 - 5x_1 + 1$ (cf. Example 4.1[3]) then the isomorphism

$$X_F \cong \left(\mathbb{T}^2 \right)^{\mathbb{Z}}, \tag{4.4}$$

sends the transformation $T_{F,(1,0)}$ to the infinite product of toral automorphisms $T_A^{\mathbb{Z}}$ where $A = \begin{bmatrix} 0 & 1 \\ -1 & 5 \end{bmatrix}$ and the transformation $T_{F,(0,1)}$ to the shift map with alphabet $\mathbb{T}^2$. The Haar measure on (4.4) is simply the infinite product of the Lebesgue measure on the torus.

It is enough to consider a set of the form

$$B = \{(\mathbf{z}_k) \in (\mathbb{T}^2)^{\mathbb{Z}} : \mathbf{z}_k \in B_\epsilon \text{ for } |k| < N\}$$

for some N and some neighbourhood of the identity B_ϵ in $\mathbb{T}^2$. Now

$$\bigcap_{0 \le k_i < m; 1 \le i \le n} T_{F, -\mathbf{k}}(B)$$

corresponds under the isomorphism (4.4) to the set

$$\{(\mathbf{z}_k) \in (\mathbb{T}^2)^{\mathbb{Z}} : \mathbf{z}_k \in B_\epsilon \text{ for } -N \le k \le N + m - 1\},$$

which has measure

$$\left(\lambda_{\mathbb{T}^2} \left(\bigcap_{j=0}^{m-1} T_A^{-j}(B_\epsilon) \right) \right)^{2N+m}$$

Now in the proof of Theorem 2.12 we showed that

$$\frac{1}{m} \log \lambda_{\mathbb{T}^2} \left(\bigcap_{j=0}^{m-1} T_A^{-j}(B_\epsilon) \right) \longrightarrow m(F)$$

where F is the characteristic polynomial of A, as $m \to \infty$. It follows that

$$h(T_F) = h(T_A) = m(F),$$

where the first entropy is that of a $\mathbb{Z}^2$-action and the second is of a $\mathbb{Z}$-action.

A non-trivial consequence of Theorem 4.2 is a dynamical proof of a result proved by Smyth as part of his characterization of polynomials whose Mahler measure vanishes. Let $F \in \mathbb{Z}[x_1^{\pm 1}, \ldots, x_n^{\pm 1}]$ be a polynomial with Newton polygon $\mathcal{C}(F)$. A face of F is the polynomial obtained by reading along a line that meets the boundary of $\mathcal{C}(F)$ in a line. Notice that each face is a polynomial of the form $x_1^{a_1} x_2^{a_2} H(x_1^{b_1} x_2^{b_2})$ for some $H \in \mathbb{Z}[x]$.

Lemma 4.4. *If $x_1^{a_1} x_2^{a_2} H(x_1^{b_1} x_2^{b_2})$ is a face of $F \in \mathbb{Z}[x_1^{\pm 1}, x_2^{\pm 1}]$, then*

$$m(H) \le m(F),$$

where $m(H)$ is a one-dimensional Mahler measure and $m(F)$ is a two-dimensional Mahler measure.

Proof. See Lind, Schmidt and Ward [LSW90, Remark 5.5] and Schmidt [Sch95, Lemma 19.6], where this is proved using the equivalent separated set definition of entropy.

4.3 Periodic Point Behaviour

Recall from Section 2.4 that if C is any $d \times d$ integer matrix, the map T_C of the torus $\mathbb{T}^d$ has $|\det(C)|$ points in its kernel (if this quantity is non-zero). We write K for the unit (multiplicative) circle $\mathbb{S}^1 = \{z \in \mathbb{C} : |z| = 1\}$. We will use several times the simple factorization

$$a^n - 1 = \prod_{j=0}^{n-1} (a - e^{2\pi i j/n}). \tag{4.5}$$

Lemma 4.5. *If $F \in \mathbb{Z}[x_1, \ldots, x_n]$ is a non-zero polynomial and $\Lambda \subset \mathbb{Z}^n$ is a finite-index subgroup, then the number of points in X_F with period Λ under the action T_F is given by*

$$|\mathrm{Per}_\Lambda(T_F)| = \prod_{\mathbf{z} \in \Omega(\Lambda)} |F(\mathbf{z})| \tag{4.6}$$

if the right-hand side is non-zero, where

$$\Omega(\Lambda) = \{\mathbf{z} \in K^n : \mathbf{z}^{\mathbf{n}} = 1 \text{ for all } \mathbf{n} \in \Lambda\}.$$

Example 4.6. [1] If $n = 1$ then this generalizes the formula in Lemma 2.14: this is a $\mathbb{Z}$-action so we must have $\Lambda = n\mathbb{Z}$ for some $n \geq 1$ and $F(x) = a_d x^d + \ldots + a_0$. According to Lemma 2.14, the number of points with period Λ is

$$|\mathrm{Per}_n(T_A)| = \left| a_d^n \prod_{k=1}^{d} (\alpha_k^n - 1) \right|$$

where $\alpha_1, \ldots, \alpha_d$ are the zeros of F. Using (4.5) we see that

$$\left| a_d^n \prod_{k=1}^{d} (\alpha_k^n - 1) \right| = \left| a_d^n \prod_{k=1}^{d} \prod_{j=0}^{n-1} (\alpha_k - e^{2\pi i j/n}) \right|$$

$$= \prod_{j=0}^{n-1} |a_d| \prod_{k=1}^{d} \left| \alpha_k - e^{2\pi i j/n} \right|$$

$$= \prod_{j=0}^{n-1} \left| F(e^{2\pi i j/n}) \right|,$$

in accordance with Lemma 4.5.

[2] The right-hand side of equation (4.6) will vanish for certain periods if there are joint unit roots in the set of common zeros of F. If $F(x_1, x_2) = 1 + x_1 + x_2$, then $F(e^{2\pi i/3}, e^{4\pi i/3}) = F(e^{4\pi i/3}, e^{2\pi i/3}) = 0$, so that (4.6) will be non-zero if and only if Λ is not contained in $3\mathbb{Z} \times 3\mathbb{Z}$.

A more extreme example is given by the polynomial

$$F(x_1, x_2) = x_1 + x_2 - 4x_1x_2 + x_1^2x_2 + x_1x_2^2;$$

in this case (4.6) vanishes for every period Λ since $F(1,1) = 0$.

Periods for which (4.6) vanishes have infinitely many periodic points; this does not arise for the dynamical systems of Chapter 2 in the ergodic case, whereas both of the polynomials considered above give rise to ergodic $\mathbb{Z}^2$-actions (see Schmidt [Sch95] for a treatment of ergodicity for higher-dimensional dynamical systems).

Proof (of Lemma 4.5 in a special case). The proof in the general case may be found in Lind, Schmidt and Ward [LSW90, Section 7]. A simple example will show how the expression in (4.6) arises.

Let $F(x_1, x_2) = 1 + 3x_1 + x_2$, and assume that $\Lambda = (k_1, 0)\mathbb{Z} \times (0, k_2)\mathbb{Z}$ is a rectangular lattice. Write $T = T_F$ and $X = X_F$. Consider the set of points which are fixed by $T_{(k_1,0)}$,

$$X_{(k_1,0)} = \{z \in X_F : z_{(a,b)} = z_{(a+k_1,b)} \text{ for all } a, b \in \mathbb{Z}\}.$$

If we define vectors

$$\mathbf{z}_k = \left(z_{(0,k)}, z_{(1,k)}, \ldots, z_{(k_1-1,k)}\right)$$

for $k \in \mathbb{Z}$ then $X_{(k_1,0)}$ is identified with the group

$$Y = \{(\mathbf{z}_k) \in (\mathbb{T}^{k_1})^{\mathbb{Z}} : \mathbf{z}_{k+1}^t = A\mathbf{z}_k^t \text{ for all } k \in \mathbb{Z}\},$$

where A is the $k_1 \times k_1$ circulant matrix

$$A = \begin{bmatrix} -3 & -1 & 0 & \cdots & \cdots & 0 \\ 0 & -3 & -1 & \cdots & \cdots & 0 \\ 0 & 0 & -3 & \cdots & \cdots & 0 \\ \vdots & & & \ddots & & \\ \vdots & & & & -3 & -1 \\ -1 & 0 & 0 & \cdots & \cdots & -3 \end{bmatrix}.$$

The eigenvalues of A are $-3 - e^{2\pi ij/k_1}$ for $j = 0, 1, \ldots, k_1 - 1$. Using the remark at the start of this section and the observation (4.5), it follows that

$$\begin{aligned} |\mathrm{Per}_\Lambda(T_F)| &= |\det(A^{k_2} - I)| \\ &= \left| \prod_{j=0}^{k_1-1} \left(\left(-3 - e^{2\pi ij/k_1} \right)^{k_2} - 1 \right) \right| \\ &= \left| \prod_{j=0}^{k_1-1} \prod_{\ell=0}^{k_2-1} \left(-3 - e^{2\pi ij/k_1} - e^{2\pi i\ell/k_2} \right) \right| \\ &= \prod_{\mathbf{z} \in \Omega(\Lambda)} |F(\mathbf{z})|. \end{aligned}$$

The notion of expansiveness used in Section 2.3 has a straightforward extension to $\mathbb{Z}^n$-actions, and the proof of Theorem 2.13 extends to give one direction of the result.

An open neighbourhood $U \subset X_F$ of the identity 0 is an expansive neighbourhood for T_F if

$$\bigcap_{\mathbf{n} \in \mathbb{Z}^n} T_F^{-\mathbf{n}}(U) = \{0\},$$

and T_F is expansive if and only if it has an expansive neighbourhood.

Theorem 4.7. *Let $F \in \mathbb{Z}[x_1, \ldots, x_n]$ be a polynomial with $F(0) \neq 0$. Then T_F is expansive if and only if there is no n-tuple $(\alpha_1, \ldots, \alpha_n) \in K^n$ with $F(\alpha_1, \ldots, \alpha_n) = 0$.*

Proof. The 'if' direction may be seen exactly as in the one-dimensional case, and this is left as an exercise.

For the reverse direction, see Schmidt [Sch95, Theorem 6.5] or argue as in Exercise 2.6: if $(\alpha_1, \ldots, \alpha_n) \in K^n$ has $F(\alpha_1, \ldots, \alpha_n) = 0$, then for any $\epsilon > 0$ the point $x \in \mathbb{T}^{\mathbb{Z}^n}$ defined by

$$x_{\mathbf{m}} = \epsilon \Re \alpha_1^{m_1} \ldots \alpha_n^{m_n}$$

lies in X_F. Since $\epsilon > 0$ was arbitrary, it follows that T_F cannot act expansively.

Exercise 4.1. Modify the proof of Theorem 2.13 to prove that if there is no n-tuple $(\alpha_1, \ldots, \alpha_n) \in K^n$ with $F(\alpha_1, \ldots, \alpha_n) = 0$ then T_F is expansive.

Example 4.8. [1] For $n = 1$ this simply reduces to Theorem 2.13.
[2] The polynomial $F(x_1, x_2) = 1 + x_1 + x_2$ has two points on K^2 where it vanishes. It follows that T_F is not expansive.
[3] The polynomial $F(x_1, x_2) = 1 + 3x_1 + x_2$ considered in Lemma 4.5 gives rise to an expansive action.

Exercise 4.2. Prove that if T_F is expansive, then T_F has finitely many periodic points for each period. Is the converse true?

Just as in Chapter 2 the expansive systems have a well-behaved growth rate of periodic points. The following result was proved in Lind, Schmidt and Ward [LSW90, Proposition 7.4]. For a period $\Lambda \subset \mathbb{Z}^n$, define

$$g(\Lambda) = \min_{0 \neq \mathbf{n} \in \Lambda} \|\mathbf{n}\|.$$

Notice that if $g(\Lambda)$ is large then the index $|\mathbb{Z}^n/\Lambda|$ is large, but not conversely.

Theorem 4.9. *If T_F is an expansive dynamical system then the growth rate of periodic points exists and equals the entropy:*

$$\lim_{g(\Lambda) \to \infty} \frac{1}{|\mathbb{Z}^n/\Lambda|} \log |\mathrm{Per}_\Lambda(T_F)| = h(T_F) = m(F).$$

Proof. By Theorem 4.7 the function $\mathbf{z} \mapsto \log|F(\mathbf{z})|$ is continuous on K^n. If $g(\Lambda) \to \infty$, then $\Omega(\Lambda)$ is a set of points on K^n which eventually meets any open set. It follows that

$$\frac{1}{|\mathbb{Z}^n/\Lambda|} \log|\mathrm{Per}_\Lambda(T_F)|$$

for a large value of $g(\Lambda)$ is simply a Riemann approximation to the integral $\int_{K^n} \log|F|$, with a continuous integrand.

Without the assumption of expansiveness, the existence of the growth rate of periodic points depends on exactly how close the zeros of F on K^n (which create logarithmic singularities in the integrand) lie to points in $\Omega(\Lambda)$ in terms of $g(\Lambda)$. For $n = 1$ this question is (more than) adequately answered by Baker's theorem (cf. Theorem 2.16[1], Lind [Lin82, Section 4]). For $n > 1$ the situation is not clear: see the discussion at the end of Schmidt [Sch95, Section 21].

Question 8. If $F \in \mathbb{Z}[x_1, \ldots, x_n]$ has the property that T_F has finitely many points of each period but T_F is not expansive, does it follow that the growth rate of periodic points exists?

Exercise 4.3. Assume that the only zeros of $F \in \mathbb{Z}[x_1, \ldots, x_n]$ on K^n are n-tuples of proper unit roots, and none of these have 1 as a component. Show that the growth rate of the number of periodic points along sequences of periods for which there are only finitely many periodic points exists and equals the entropy.

Question 9. Is there an analogue of Theorem 2.16[2] for $\mathbb{Z}^n$-actions?

The number of periodic points of each period is a topological invariant for dynamical systems; in [War98b] the strength of this invariant is examined for a simple class of algebraic examples of $\mathbb{Z}^2$-actions.

4.4 Dynamical Zeta Functions

The dynamical zeta function for a single transformation was described in Exercise 2.7. Some progress has been made on defining a natural zeta function for higher-dimensional actions by Lind [Lin96] and the exercises below are taken from his paper.

Definition 4.10. Let T be an action of $\mathbb{Z}^n$ with the property that $\mathrm{Per}_\Lambda(T)$ is finite for all periods $\Lambda \subset \mathbb{Z}^n$. Then the *dynamical zeta function* of T is

$$\zeta_T(z) = \exp \sum_\Lambda \frac{|\mathrm{Per}_\Lambda(T)|}{|\mathbb{Z}^n/\Lambda|} \cdot z^{|\mathbb{Z}^n/\Lambda|}, \tag{4.7}$$

where the summation is taken over all finite-index subgroups of $\mathbb{Z}^n$.

Notice that this reduces to the definition in Exercise 2.7 when $n = 1$. There are considerable difficulties attached to the expression (4.7), not the least of which is the need to parametrize the finite-index subgroups of $\mathbb{Z}^n$ in a convenient way.

Exercise 4.4. Show that the subgroups of finite index in $\mathbb{Z}^2$ are faithfully parametrized by the set of matrices

$$\left\{ \begin{bmatrix} a & b \\ 0 & c \end{bmatrix} : a, c \geq, 0 \leq b \leq a - 1 \right\},$$

under the correspondence sending the matrix A to the subgroup $\Lambda(A) = \{Ax \in \mathbb{Z}^2\}$ (writing x as a column vector).

Exercise 4.5. [1] Find an expression for the zeta function of the trivial $\mathbb{Z}^2$-action on a singleton (cf. Lind [Lin96, Example 3.1]).
[2] Find an expression for the zeta function of the action T_F on X_F where $F(x_1, x_2) = k$ is a constant polynomial (cf. Lind [Lin96, Example 3.2]).

Exercise 4.6. Define an action T as follows. The generator $T_{(0,1)}$ is the one-dimensional action defined by the constant polynomial $k \in \mathbb{Z}[x]$ (cf. Section 2.3); the generator $T_{(1,0)}$ is the identity map. Thus the underlying group is

$$X = \prod_{\mathbb{Z}} \{0, 1 \ldots, |k| - 1\}$$

and the action is defined by $T_{(n,m)} = \sigma^m$, where σ is the left shift on X.
[1] Compute the topological entropy of this action.
[2] Compute the zeta function for this action, and locate the smallest real pole (cf. Lind [Lin96, Example 3.3]).

There are several interesting open problems in Lind's paper [Lin96], from which we single out the following central problem.

Question 10. Is it true that the dynamical zeta function of T_F is meromorphic in the disc $|z| < e^{-h(T_F)}$ and has the circle $|z| = e^{-h(T_F)}$ as natural boundary? (cf. Lind [Lin96, Conjecture 7.1]).

5. Elliptic Heights

5.1 Elliptic Curves

An elliptic curve is a mathematical object with many facets: geometric, analytic, number-theoretic. Rigorous treatments may be found in the books by Husemöller [Hus87], Koblitz [Kob84] and Silverman [Sil86], [Sil94].

Let K be a fixed field with characteristic not equal to 2 or 3. The study of elliptic curves begins for our purposes with the set of solutions to the affine equation

$$y^2 = x^3 + ax + b, \tag{5.1}$$

where $a, b \in K$ and the discriminant $\Delta = 4a^3 + 27b^2$ is non-zero.

Exercise 5.1. [1] If the right-hand side of (5.1) factorizes as

$$x^3 + ax + b = (x - \alpha_1)(x - \alpha_2)(x - \alpha_3)$$

then show that the discriminant of the curve is given by

$$\Delta = - \left((\alpha_1 - \alpha_2)(\alpha_1 - \alpha_3)(\alpha_2 - \alpha_3) \right)^2.$$

Deduce that the discriminant of the curve vanishes if and only if the equation $x^3 + ax + b = 0$ has a multiple root.
[2] What happens to the discriminant in characteristic 2 or 3?

Remark 5.1. In our treatment, the discriminant Δ of the curve differs in sign from the discriminant of the polynomial $x^3 + ax + b$ (cf. Remark 1.5). The only properties of the discriminant we will need relate to vanishing and divisibility.

More correctly, an elliptic curve is a projective object.

Definition 5.2. Let K be any field, and $1 \leq N \in \mathbb{N}$. Then the *projective space* of dimension N over K is

$$\mathbb{P}^N(K) = \left\{ (x_0, \ldots, x_N) \neq (0, \ldots, 0) \in K^{N+1} \right\} / \sim \tag{5.2}$$

where $\mathbf{x} \sim \mathbf{x}'$ if and only of $\mathbf{x} = \lambda \mathbf{x}'$ for some $\lambda \in K \backslash \{0\}$. Use homogeneous coordinates

$$[\mathbf{x}] = [x_0, \ldots, x_n]$$

to denote the equivalence class of a point.

Remark 5.3. [1] The word 'dimension' is used here in the sense of algebraic geometry; $\mathbb{P}^N(K)$ is not a vector space and so has no dimension in the linear sense.

[2] We will be interested in the case $N = 2$. A subset of $\mathbb{P}^2(K)$ is an *algebraic curve* if it comprises the set of zeros of a non-zero homogeneous polynomial $f(x, y, z)$.

Example 5.4. The set of zeros of the polynomial

$$y^2 z - x^3 - a x z^2 - b z^3 \tag{5.3}$$

forms an algebraic curve in $\mathbb{P}^2(K)$. This curve has two pieces:

(i) if $z = 0$ then $x = 0$ so $y \neq 0$; thus there is exactly one projective point on (5.3) with $z = 0$, namely

$$[0, 1, 0].$$

(ii) if $z \neq 0$ then we may take it to be 1, so all the points

$$[x, y, 1]$$

with $y^2 = x^3 + ax + b$ lie on the curve.

This curve (with $4a^3 + 27b^2 \neq 0$) is an *elliptic curve*.

Exercise 5.2. The specific form of equation (5.1) with no second-order term in x is less special than at first appears. In general, an elliptic curve is the set of zeros of a cubic equation in $\mathbb{P}^2(K)$ with only one point at infinity, $[0, 1, 0]$. Following Silverman [Sil86, Section III.1], write the general equation of such a projective curve as

$$Y^2 Z + a_1 XYZ + a_3 Y Z^2 = X^3 + a_2 X^2 Z + a_4 X Z^2 + a_6 Z^3$$

(after scaling X and Y). Write the the affine equation for the affine part of the curve using the variables $x = X/Z$ and $y = Y/Z$. Show that replacing y by $\frac{1}{2}(y - a_1 x - a_3)$ gives a curve of the form

$$y^2 = \text{cubic in } x.$$

Show that the replacing x by $(x - c)/36$ and y by $y/108$ where

$$c = 3(a_1^2 a_6 + 4a_2 a_6 - a_1 a_3 a_4 + a_2 a_3^2 - a_4^2)$$

eliminates the second-order term in x, giving an equation in the form (5.1).

Since the substitutions are all rational functions with coefficients in K, there is a bijection between the points on the first curve and the points on the last curve. This justifies the use of the standard form (5.1).

The rough picture is therefore as follows. The projective curve defined by (5.3) contains a copy of the affine curve defined by (5.1) together with one extra point, obtained by intersecting the affine curve with the 'line at infinity'. This extra point $[0, 1, 0]$ corresponds to some sort of limit of points (x, y) with $y/x \to \infty$. This single point 'at infinity', usually denoted 0, will turn out to be a key player in the structure of the curve defined by (5.3): it will be the identity element for the group structure found.

Passing to the projective setting is also much simpler because Bezout's theorem (see Hartshorne [Har77, Corollary I.7.8]) holds there: if K is algebraically closed, then a curve of degree d_1 and a curve of degree d_2 in $\mathbb{P}^2(K)$ will intersect in exactly $d_1 d_2$ points (counted with multiplicity).

Exercise 5.3. A line in $\mathbb{P}^2(K)$ is given by an equation

$$Ax + By + Cz = D$$

with $A, B, C, D \in K$. Prove that any pair of lines meet in exactly one point.

Definition 5.5. Write L/K to mean that L is an extension field of K. Write

$$E : y^2 = x^3 + ax + b \text{ with } a, b \in K$$

to mean that the projective elliptic curve defined by the homogeneous equation (5.3) is called E, and write $E(L)$ for the set of solutions $[x, y, z] \in \mathbb{P}^2(L)$ to (5.3). We will say that E is *defined over* K in this case.

Exercise 5.4. Let $E : y^2 = x^3 - x$. Show that $E(\mathbb{Q})$ has four points.

For any L/K, define a binary operation $+$ on $E(L)$ as follows. If P and Q are points on $E(L)$, then the line through P and Q meets $E(L)$ in exactly one further point (x, y) say. The reflection $R = (x, -y)$ of (x, y) in the x-axis is then defined to be $P + Q$ (see Figure 5.1).

Theorem 5.6. *The set $E(L)$ with binary operation $+$ forms an abelian group.*

Over an arbitrary field, the proof is non-trivial. The difficult step is to check that $+$ is associative, and there are (at least) four approaches:

- Function-theoretic: this is the approach we take in Chapter 6, when the field $L = K = \mathbb{C}$. The set of $\mathbb{C}$-points turns out to be canonically in bijection with the points of a certain group (in fact a quotient group of $\mathbb{C}$ itself). Under this bijection, which is given by explicit complex functions, the group operation corresponds to the geometric addition just defined. See Theorem 6.4 for the details.
- Geometric (see for example Cassels [Cas91, Chapter 7]); a good exercise is to try and draw a convincing picture of what associativity means over the reals.

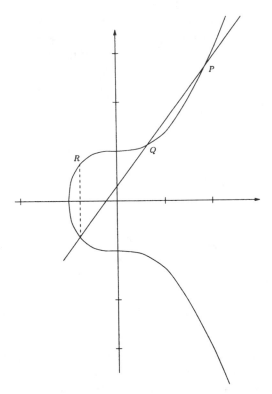

Fig. 5.1. The binary operation on $E(\mathbb{R})$

- Muscular (see Silverman [Sil86, Section III.2]): here one explicitly computes functions $\phi, \psi : L^4 \to L$ with the property that if P and Q are points on $E(L)$ then $P + Q = (\phi(P,Q), \psi(P,Q))$ and checks that the resulting operation is associative. Explicit formulæ are given in Section 5.2 below and the group properties can – in principle – be checked by hand.
- Sophisticated: using divisors and the Riemann–Roch theorem (see Silverman [Sil86, Section III.3]).

We simply motivate some of the ingredients for the real case using pictures. A complete treatment may be found in Silverman [Sil86]; modern algebraic geometry has an explanation of why an elliptic curve over an arbitrary base field has a group structure.

The case $P = Q$ requires a notion of 'tangency' (which can be defined for curves over any field using order of vanishing), and then $2P$ is obtained by reflecting the unique other point of intersection of the line tangent to the curve at P in the x-axis.

The identity element is the point at infinity, which in the affine pictures may be thought of as reached by moving infinitely far up (or down) the y-axis. In Figure 5.2, a sequence of elements is shown converging to 0, the point at infinity. When each element is added to P, the resulting sequence is converging to P.

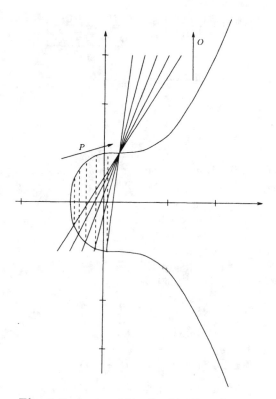

Fig. 5.2. Approaching the identity element

Finally, additive inverses are easily seen to be given by reflection in the x-axis, so if $P = (x, y)$ then $-P = (x, -y)$.

Exercise 5.5. Three points P, Q, R on $E(\mathbb{R})$ are collinear if and only if

$$P + Q + R = 0$$

(cf. Theorem 6.4[3]).

5.2 Explicit Formulæ

Given points $P_1 = (x_1, y_1)$ and $P_2 = (x_2, y_2)$, explicit formulæ may be found for x_3 and y_3, where $P_1 + P_2 = (x_3, y_3)$.

Case I: If $x_1 \neq x_2$, then the line joining P_1 to P_2 has equation

$$\frac{y - y_1}{x - x_1} = \frac{y_2 - y_1}{x_2 - x_1},$$

so

$$y = \underbrace{\left(\frac{y_2 - y_1}{x_2 - x_1}\right)}_{\alpha} x + \underbrace{\left(\frac{x_2 y_1 - x_1 y_2}{x_2 - x_1}\right)}_{\beta}.$$

Substituting this into the equation

$$y^2 = x^3 + ax + b$$

for the curve gives

$$(ax + \beta)^2 = x^3 + ax + b,$$

whose roots are the 3 (by Bezout's theorem) points of intersection. Using the sum of roots formula,

$$x_1 + x_2 + x_3 = \alpha^2,$$

so

$$x_3 = \alpha^2 - x_1 - x_2 = \left(\frac{y_2 - y_1}{x_2 - x_1}\right)^2 - x_1 - x_2.$$

Reflecting in the x-axis gives P_3, so

$$y_3 = -\alpha x_3 - \beta = -\left(\frac{y_2 - y_1}{x_2 - x_1}\right) x_3 - \left(\frac{x_2 y_1 - x_1 y_2}{x_2 - x_1}\right) \quad \text{or } y_3 = \alpha(x_1 - x_3) - y_1.$$

Case II: If $x_1 = x_2$ and $y_1 = y_2$, then we must take the tangent to the curve. By implicit differentiation of $y^2 = x^3 + ax + b$ we get

$$\alpha = \frac{3x_1^2 + a}{2y_1},$$

hence

$$x_3 = \left(\frac{3x_1^2 + a}{2y_1}\right)^2 - 2x_1, \quad y_3 = \left(\frac{3x_1^2 + a}{2y_1}\right)\left(\left(\frac{3x_1^2 + a}{2y_1}\right)^2 - 3x_1\right) + y_1.$$

or $y_3 = \alpha(x_1 - x_3) - y_1$.
Case III: If $x_1 = x_2$ and $y_1 = -y_2$ then $P_1 = -P_2$ and so P_3 is the point at infinity.

Notice that all the formulæ are rational functions: so if $P_1, P_2 \in E(L)$ then $P_1 + P_2 \in E(L)$, which means that the group operation is well-defined.

Example 5.7. Let $E : y^2 = x^3 + 1$, and let $P = (2, 3)$. Using the formulae, we find

$$2P = (0, 1); 3P = 2P + P = (-1, 0); 4P = 3P + P = (0, -1) = -2P.$$

It follows that $6P = 0$ (so P is a torsion point), and since $P, 2P, 3P \neq 0$ the point P has order six (see Figure 5.3).

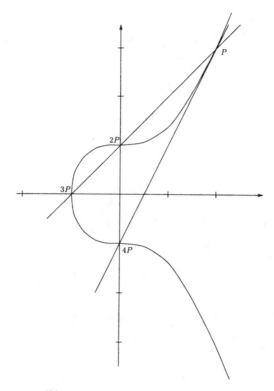

Fig. 5.3. The point P has order 6

Exercise 5.6. Find the order of the point $(3, 8)$ on the elliptic curve

$$y^2 = x^3 - 43x + 166.$$

Exercise 5.7. Suppose that K denotes any field with characteristic not equal to 2 or 3, and $E : y^2 = x^3 + ax + b$ $(a, b \in K)$. Prove that $P = (x, y)$ has order 2 if and only if $y = 0$.

Exercise 5.8. Suppose $1 \le n \in \mathbb{N}$ and consider the elliptic curve

$$y^2 = x^3 - n^2 x.$$

Prove that there are only two real points of order 3. Mark these points on a graph of the curve (they are points of inflexion). It may be interesting to look at Example 5.14.

Exercise 5.9. Use your graph from Exercise 5.8 to show that the subgroup of real points on $y^2 = x^3 - n^2 x$ with order dividing 4 is $C_2 \times C_4$.

Exercise 5.14 below describes a useful result which allows all the rational torsion points on an integral elliptic curve to be effectively determined.

When $K = \mathbb{Q}$, there are not many possibilities for the possible orders of torsion elements in $E(\mathbb{Q})$. For example, in the next section we show that there are no points of order 11 (assuming a non-trivial result from Diophantine equations).

Theorem 5.8. *There are no integral points on $E : y^2 = x^3 + 1$ except those in the group generated by $P = (2, 3)$.*

Proof (sketch proof). Let $\omega = e^{\pi i/3}$ be a primitive sixth root of unity. Then the minimal equation for ω over $\mathbb{Q}$ is

$$\omega^2 = \omega - 1,$$

so $K = \mathbb{Q}(\omega)$ is a degree 2 extension of $\mathbb{Q}$. The ring of algebraic integers in K is $\mathbb{Z}[\omega]$,

$$
\begin{array}{ccc}
\mathbb{Q}(\omega) & \!\!\!\!\!\! \text{---} \!\!\!\!\!\! & \mathbb{Z}[\omega] \\
| & & | \\
\mathbb{Q} & \!\!\!\!\!\! \text{---} \!\!\!\!\!\! & \mathbb{Z}
\end{array}
$$

and this ring is a UFD.

Factorizing the equation in $\mathbb{Z}[\omega]$ gives

$$y^2 = (x - \omega)(x - \overline{\omega})(x + 1). \tag{5.4}$$

If an integer d divides $(x - \omega)$ and $(x - \overline{\omega})$ then it must also divide $(\omega - \overline{\omega}) = \sqrt{-3}$, which is a prime in $\mathbb{Z}[\omega]$. By unique factorization, since the left-hand side of (5.4) is a square, we must have

$$x - \omega = \omega^i (a + b\omega)^2 \text{ or } \omega^i \sqrt{-3}(a + b\omega)^2$$

with $a, b \in \mathbb{Z}$ and $i = 1, \ldots, 6$. Checking the restrictions on $a, b \in \mathbb{Z}$ eventually gives the only possible values for x as $0, -1$, or 2.

Exercise 5.10. Prove that the curve $y^2 = x^3 - 1$ has only one integral point.

If the equation for an elliptic curve splits over a field whose ring of integers is a UFD, one might imagine that similar methods will show there are always going to be only finitely many integral points. In general one has to deal with the non-UFD case as well, and this was done by Siegel: see [Sie66] for the original paper.

Theorem 5.9. [SIEGEL'S THEOREM] *If E is an elliptic curve defined over $\mathbb{Q}$, then E has only finitely many integral points.*

5.3 Torsion Points of Order 11

A historically very important and fruitful question has been the following: if E is an elliptic curve, what are the torsion elements in the abelian group $E(\mathbb{Q})$? There are some very sophisticated results and problems in this direction – see Ogg [Ogg71]. Mazur has proved that the only possible orders for rational torsion are $1, \ldots, 10$ or 12, and indeed the torsion subgroup can be only one of finitely many possibilities (see Husemöller [Hus87, Chapter 1] and Silverman [Sil86, Theorem 7.5] for the details and related background). As an illustration of the constraints that can appear – and to practice working with elliptic curves – we present the following argument, which reduces the existence of points of order 11 to a (non-trivial) Diophantine problem. The presentation follows Ogg's modernization of the original paper of Billing and Mahler (see [BM40] and [Ogg71]) and was presented in a course taught by A. Silverberg at the Ohio State University in the Spring of 1992.

Theorem 5.10. *If E is an elliptic curve defined over $\mathbb{Q}$, then $E(\mathbb{Q})$ has no point of order 11.*

Proof. Assume that P is a point in $E(\mathbb{Q})$ with order 11. Then no three elements of $\{0, P, 3P, 4P\}$ can be collinear because if A, B, C are collinear then $A + B + C = 0$.

It follows that there is a non-singular linear map on $\mathbb{P}^2(\mathbb{Q})$ sending

$$0 \to [0, 1, 0], P \to [1, 0, 0], 3P \to [0, 0, 1] \text{ and } 4P \to [1, 1, 1].$$

To see this, notice that of the four points

$$[0, 1, 0], [1, 0, 0], [0, 0, 1], [1, 1, 1]$$

no three are collinear. Given any four points with homogeneous coordinates

$$\mathbf{v}_1, \mathbf{v}_2, \mathbf{v}_3, \mathbf{v}_4,$$

the matrix $M = [a\mathbf{v}_1^t | b\mathbf{v}_2^t | c\mathbf{v}_3^t]$ will, for any $a, b, c \neq 0$, send

$$[1, 0, 0] \to \mathbf{v}_1$$
$$[0, 1, 0] \to \mathbf{v}_2$$
$$[0, 0, 1] \to \mathbf{v}_3$$
$$[1, 1, 1] \to a\mathbf{v}_1 + b\mathbf{v}_2 + c\mathbf{v}_3.$$

The equation $a\mathbf{v}_1 + b\mathbf{v}_2 + c\mathbf{v}_3 = \mathbf{v}_4$ has a unique solution with a, b, c all non-zero by the non-collinearity assumption. Thus, by applying a change of variables in $\mathbb{P}^2(\mathbb{Q})$ we may assume that $0 = [0, 1, 0], P = [1, 0, 0], 3P = [0, 0, 1]$ and $4P = [1, 1, 1]$.

Now let $5P = [x_1, x_2, x_3]$. Then if ℓ_1 is the line through $5P$ and 0, and ℓ_2 is the line through $4P$ and P, $-5P \in \ell_1 \cap \ell_2$. Thus

$$r[0, 1, 0] + s[x_1, x_2, x_3] = t[1, 0, 0] + w[1, 1, 1],$$

so

$$[sx_1, sx_2 + r, sx_3] = [t + w, w, w].$$

Hence,

$$sx_1 = t + w; \ sx_2 + r = w; \ sx_3 = w.$$

If $s = 0$ then $P = 0$, which is impossible, so we may put $s = 1$. Then $r = x_3 - x_2$, and so

$$-5P = [x_1, x_3, x_3].$$

Similar arguments show that

$$-4P = [1, 0, 1],$$
$$-P = [x_1 - x_3, x_2, 0],$$
$$-3P = [0, x_3 - x_1 + x_2, x_3 - x_1],$$
$$2P = [x_1 x_3 - x_1^2 + x_1 x_2, x_3^2 - x_1 x_3 + x_2 x_3, x_3^2 - x_1 x_3].$$

Since $11P = 0$, the points $5P, 4P, 2P$ are collinear. It follows that

$$x_3^3 - x_1^2 x_2 + x_1^2 x_3 + x_1 x_2^2 - 2x_1 x_3^2 = 0. \tag{5.5}$$

The only rational solutions to (5.5) are

$$[1, 1, 1], [0, 1, 0], [1, 0, 0], [1, 0, 1], [1, 1, 0].$$

This is proved in Billing and Mahler [BM40, Lemma 2].

Now $5P$ must correspond to one of the above possibilities. It cannot be $[1, 1, 1]$ because this is $4P$ and $P \neq 0$; it cannot be $[0, 1, 0]$ because this is 0 and $10P \neq 0$; it cannot be $[1, 0, 0]$ because this is P and $4P \neq 0$; it cannot be $[1, 0, 1]$ because this is $-4P$ and $9P \neq 0$; it cannot be $[1, 1, 0]$ because this is $-P$ and $6P \neq 0$.

The contradiction proves there can be no such point P.

5.4 Rational Points

Our main interest is with rational points on elliptic curves defined over $\mathbb{Q}$.

Example 5.11. Consider the curve $y^2 = x^3 - 36x$. Let $P = \left(\frac{25}{4}, \frac{35}{8}\right)$. Then P is a rational point of infinite order, and we compute that

$$2P = \left(\frac{1\,442\,401}{19\,600}, -\frac{1\,726\,556\,399}{2\,744\,000}\right)$$

$$4P = \left(\frac{4\,386\,303\,618\,090\,112\,563\,849\,601}{233\,710\,164\,715\,943\,220\,558\,400}, -\frac{870\,369\,109\,085\,580\,828\,275\,935\,650\,626\,254\,401}{11\,298\,385\,812\,463\,619\,737\,216\,684\,496\,448\,000}\right).$$

This example is related to the 'congruent number problem'. A natural number n is said to be congruent if it arises as the area of a right triangle with rational sides. For example, 6 is a congruent number since it is the area of the $3, 4, 5$ triangle. The elliptic curve $y^2 = x^3 - 36x$ allows other rational right triangles with area 6 to be computed: using the points above, one finds the right-angled triangles with sides

$$\left(\frac{120}{7}, \frac{7}{10}, \frac{1201}{70} \right)$$

and

$$\left(\frac{2\,017\,680}{1\,437\,599}, \frac{1\,437\,599}{168\,140}, \frac{2\,094\,350\,404\,801}{241\,717\,895\,860} \right),$$

each of which has area 6. For a thorough discussion of the congruent number problem and its connection to elliptic curves, see the book by Koblitz [Kob84].

The arithmetic complexity of the rational points seems to grow enormously, and we want to quantify this growth in complexity. To do this, we will use a naïve version of 'height' on elliptic curves over the rationals, which allows us to measure how complicated rational points become under maps like $P \mapsto 2P$. This notion of height was introduced by Mordell [Mor22] with the specific aim of proving the following theorem. The conjecture that this should be so may be attributed to Poincaré [Poi01].

Theorem 5.12. [MORDELL'S THEOREM] *Let E denote an elliptic curve defined over $\mathbb{Q}$. Then $E(\mathbb{Q})$ is a finitely generated abelian group.*

Notice this is not a consequence of any general ideas about 'dimension': for example, the multiplicative group $\mathbb{Q}^*$ that sits inside the projective line $\mathbb{P}^1(\mathbb{Q})$ is not finitely generated. The underlying *genus* of the curve (1 for an elliptic curve) is responsible. Indeed, for curves of genus ≥ 2 there are only finitely many rational points, a result due to Faltings [Fal91].

5.5 Heights on Elliptic Curves

Given a rational point $P = (\frac{M}{N}, *)$, where M and N are coprime integers, define the *naïve height* to be

$$H(P) = \begin{cases} \max\{|M|, |N|\} & \text{if } P \neq 0, \\ 1 & \text{if } P = 0. \end{cases}$$

Write $P = (x(P), y(P))$ for the coordinates of a point P. As usual, we associate the logarithmic height $h(P) = \log H(P)$. Notice that $h(P)$ is a Mahler measure: $\log \max\{|M|, |N|\} = m(Nx - M)$.

Looking at Example 5.11, it appears that the length (number of digits) in numerator and denominator roughly quadruples each time the point is doubled. This is a manifestation of a general phenomenon, the *duplication formula*.

Theorem 5.13. *Let E denote an elliptic curve defined over the rationals, and let P be a point in $E(\mathbb{Q})$. Then*

$$h(2P) = 4h(P) + O(1), \tag{5.6}$$

where the implied constant in O depends on E but not on the point P. In multiplicative notation,

$$H(P)^4 \ll H(P) \ll H(P)^4.$$

Example 5.14. Consider the curve $E : y^2 = x^3 - n^2x$, with $1 \leq n \in \mathbb{N}$. If $P = (x, y)$ is a point on E, then

$$x(2P) = \left(\frac{x^2 + n^2}{2y}\right)^2 = \frac{(x^2 + n^2)^2}{4(x^3 - n^2x)},$$

so if $x = \frac{M}{N}$ in lowest terms, then

$$x(2P) = \frac{(M^2 + n^2N^2)^2}{4N^2(M^2 - n^2N^2)}. \tag{5.7}$$

It may be checked that any cancellation in (5.7) is bounded: explicitly, if d divides both numerator and denominator, then $d|4n^2$. Now examining the cases $|M| \geq |N|$ and $|M| < |N|$ separately shows that

$$\max\{|M^2 + n^2N^2|^2, |N^2(M^2 - n^2N^2)|\}$$

is commensurate with

$$\max\{|M|^4, |N|^4\} = \max\{|M|, |N|\}^4,$$

and the duplication formula (5.6) follows.

Exercise 5.11. Verify Theorem 5.13 for the curve $y^2 = x^3 + 6$.

To prove the duplication formula, we are going to use some ideas from algebraic geometry.

Definition 5.15. For $1 \leq M, N \in \mathbb{N}$, a map

$$f : \mathbb{P}^N(\mathbb{Q}) \longrightarrow \mathbb{P}^M(\mathbb{Q})$$

is a *morphism* if

$$f([\mathbf{x}]) = f([x_0, \ldots, x_N]) = [f_0([\mathbf{x}]), \ldots, f_M([\mathbf{x}])]$$

where the f_j, $0 \leq j \leq M$ are homogeneous polynomials of the same degree, and the only common zero is 0. If this common degree is d, then f is a *morphism of degree d.*

We also need to extend the notion of height to $\mathbb{P}^n(\mathbb{Q})$. Given a point $[\mathbf{x}] \in \mathbb{P}^n(\mathbb{Q})$, choose representatives in such a way that

$$[\mathbf{x}] = [x_0, \ldots, x_n]$$

with $x_i \in \mathbb{Z}$ and

$$\mathrm{hcf}(x_0, \ldots, x_n) = 1.$$

Then define

$$H : \mathbb{P}^n(\mathbb{Q}) \to \mathbb{R}$$

by

$$H([\mathbf{x}]) = \max\{|x_i|\}. \tag{5.8}$$

Notice that this is compatible with the affine height: if $P = (x, y) \in E(\mathbb{Q})$ is an affine point, then $H(P) = H(x, 1)$ where $(x, 1) \in \mathbb{P}^1(\mathbb{Q})$.

Lemma 5.16. *Suppose $1 \leq N, M$ and $f : \mathbb{P}^N(\mathbb{Q}) \to \mathbb{P}^M(\mathbb{Q})$ is a morphism of degree d. Then*

$$H([\mathbf{x}])^d \ll H(f([\mathbf{x}])) \ll H([\mathbf{x}])^d.$$

Proof. Write $f([\mathbf{x}]) = [f_0(\mathbf{x}), \ldots, f_M(\mathbf{x})]$ where $[\mathbf{x}] = [x_0, \ldots, x_N] \in \mathbb{P}^N(\mathbb{Q})$. Assume without loss of generality that each x_i is an integer. Since each f_i is a homogeneous polynomial of degree d,

$$f_i(\mathbf{x}) = \sum c_{\mathbf{e}} x_0^{e_0} \ldots x_N^{e_N},$$

with the coefficients $c_{\mathbf{e}} \in \mathbb{Q}$, $e_i \in \mathbb{N}$, $e_0 + \ldots + e_N = d$, and only finitely many $c_{\mathbf{e}}$ non-zero. It follows that

$$|f_i(\mathbf{x})| \leq C \cdot (\max\{|x_i|\})^d,$$

so there is a similar bound for $\max\{|f_i(\mathbf{x})|\}$. To find the height, notice that the only possible denominators that need to be cleared come from the coefficients of the f_i's, which is a bounded quantity. It follows that there is an upper bound for the height of the form

$$C \cdot H(\mathbf{x})^d.$$

To get the lower bound, we use Hilbert's Nullstellensatz (see Appendix A, Section A.3): there exists $e \in \mathbb{N}$ and polynomials $g_{ij} \in \mathbb{Q}[\mathbf{x}]$ such that

$$x_0^e \;=\; g_{00}(\mathbf{x})f_0(\mathbf{x}) \;+\; \cdots \;+\; g_{0N}(\mathbf{x})f_N(\mathbf{x})$$

$$\vdots$$

$$x_N^e \;=\; g_{N0}(\mathbf{x})f_0(\mathbf{x}) \;+\; \cdots \;+\; g_{NN}(\mathbf{x})f_N(\mathbf{x}).$$

The g_{ij}s can be taken to be homogeneous of degree $e - d$.

Since each g_{ij} is homogeneous of degree $e - d$,

$$|g_{ij}(\mathbf{x})| \ll (\max\{|x_i|\})^{e-d}.$$

On the other hand,

$$x_j^e = g_{j0}(\mathbf{x}) f_0(\mathbf{x}) + \ldots + g_{jN}(\mathbf{x}) f_N(\mathbf{x})$$

for $j = 0, \ldots, N$, so

$$(\max\{|x_j|\})^{e-d} \max\{|f_0|, \ldots, |f_N|\} \gg (\max\{|x_j|\})^e.$$

It follows that

$$\max\{|f_0|, \ldots, |f_N|\} \gg (\max\{|x_j|\})^d,$$

and since the only possible denominators are those arising from the coefficients of the f_i the lower bound is proved.

Example 5.17. To see that $e > d$ occurs in the Nullstellensatz, define

$$f : \mathbb{P}^1(\mathbb{Q}) \to \mathbb{P}^1(\mathbb{Q})$$

by

$$f : [x_0, x_1] \mapsto [x_0^2, (x_0 + x_1)^2] = [f_0(x_0, x_1), f_1(x_0, x_1)].$$

Then f is a morphism of degree 2. Now $x_0^2 = 1 \cdot f_0$, but there are no rational polynomials A, B for which $x_1^2 = A \cdot f_0 + B \cdot f_1$. However,

$$x_0^3 = x_0 \cdot f_0$$
$$x_1^3 = (2x_0 + 3x_1) \cdot f_0 + (-2x_0 - x_1) \cdot f_1.$$

Exercise 5.12. Using the explicit formulæ from Example 5.14 prove that $[x(P), 1] \mapsto [x(2P), 1]$ is a morphism of degree 4 for the curve $y^2 = x^3 - n^2 x$.
Do the same for $y^2 = x^3 + A$.

Proof (of Theorem 5.13). We need to show that the map

$$[x, 1] \mapsto [x(2P), 1]$$

on $\mathbb{P}^1(\mathbb{Q})$ is a morphism of degree 4.
Let $y^2 = x^3 + ax + b$ and $P = (x, y)$. Then

$$\begin{aligned}
x(2P) &= \left(\frac{3x^2 + a}{2y}\right)^2 - 2x \\
&= \frac{(3x^2 + a)^2}{4y^2} - 2x \\
&= \frac{9x^4 + 6x^2 a + a^2}{4(x^3 + ax + b)} - 2x \\
&= \frac{x^4 - 2x^2 a - 8xb + a^2}{4(x^3 + ax + b)}.
\end{aligned}$$

Write $x = \frac{x_0}{x_1} \in \mathbb{Q}$ (in lowest terms as usual). Then, writing $x(2P) = \frac{f_0(x_0,x_1)}{f_1(x_0,x_1)}$ and dropping a factor of 4,

$$f_0(x_0, x_1) = x_0^4 - 2x_0^2 x_1^2 a - 8x_0 x_1^3 b + a^2 x_1^4$$
$$f_1(x_0, x_1) = x_0^3 x_1 + a x_0 x_1^3 + b x_1^4.$$

To show that these define a morphism of degree 4 it only remains to show that the only common zero of f_0 and f_1 is $(0,0)$. If $f_0 = f_1 = 0$ and $x_1 = 0$ then $x_0 = 0$. Assume $x_1 \neq 0$. Then we may assume that $x_1 = 1$ and $x_0 = x$. We now need to show that

$$f = x^4 - 2x^2 a - 8xb + a^2$$
$$g = x^3 + ax + b$$

cannot have a common zero. The resultant of f and g with respect to x is

$$\det \begin{bmatrix} 1 & 0 & -2a & -8b & a^2 & 0 & 0 \\ 0 & 1 & 0 & -2a & -8b & a^2 & 0 \\ 0 & 0 & 1 & 0 & -2a & -8b & a^2 \\ 1 & 0 & a & b & 0 & 0 & 0 \\ 0 & 1 & 0 & a & b & 0 & 0 \\ 0 & 0 & 1 & 0 & a & b & 0 \\ 0 & 0 & 0 & 1 & 0 & a & b \end{bmatrix} = \left(4a^3 + 27b^2 \right)^2. \tag{5.9}$$

By assumption, this is non-zero.

Exercise 5.13. Give a different proof that the polynomials

$$f = x^4 - 2x^2 a - 8xb + a^2$$
$$g = x^3 + ax + b$$

cannot have a common zero by using the Euclidean algorithm to show that the greatest common divisor of f and g is $4a^3 + 27b^2$.

Exercise 5.14. Let $E : y^2 = x^3 + ax + b$ with $a, b \in \mathbb{Z}$ denote an elliptic curve. Using arguments from p-adic analysis (cf. Section 5.8 below), it can be shown that any non-zero torsion point $Q \in E(\mathbb{Q})$ must have $x(Q)$ and $y(Q)$ integral. Assuming this, prove that $y(Q) = 0$ or $y(Q)^2$ divides the discriminant $4a^3 + 27b^2$ for any rational torsion point.

The statement about integrality of torsion points in the last exercise is due to Lutz [Lut37] and Nagell [Nag35]. Accessible proofs are in Cassels [Cas91, Chapter 12], Husemöller [Hus87, Chapter 5, Section 6] and Silverman [Sil86, Chapter VIII, Section 7]. The characterization of all torsion points on the curves $y^2 = x^3 + ax$ (for a integral and not divisible by a fourth power) and $y^2 = x^3 + b$ (for b integral and not divisible by a sixth power) is given in Cassels (Exercise to Chapter 12, *ibid.*).

Example 5.18. In Exercise 5.6 we showed that the point $P = (3, 8)$ has order 7 on the elliptic curve

$$y^2 = x^3 - 43x + 166.$$

Using Exercise 5.14 (and some patient checking) we can show that there are no rational torsion points beside those generated by P.

5.6 Mordell's Theorem

On the way to proving Mordell's theorem we need the following lemma.

Lemma 5.19. [1] *If $P_0 \neq 0$ is a point in $E(\mathbb{Q})$, then there is a constant $c_1 = c_1(E, P_0) > 0$ such that*

$$h(P + P_0) < 2h(P) + c_1.$$

[2] *Given $h_0 > 0$, there are only finitely many points $P \in E(\mathbb{Q})$ with*

$$h(P) < h_0.$$

Proof. Part [2] is clear: there are only finitely many rationals $\frac{m}{n}$ (in lowest terms) for which $\log \max\{|m|, |n|\} < h_0$.

To prove part [1], write $P = (x, y)$ and $P_0 = (x_0, y_0)$. From the equation $y^2 = x^3 + ax + b$ we may write in lowest terms

$$x = \frac{r}{t^2}, \quad y = \frac{s}{t^3}, \quad x_0 = \frac{r_0}{t_0^2}, \quad y_0 = \frac{s_0}{t_0^3}$$

with r, s, t, r_0, s_0, t_0 all integers. Then

$$
\begin{aligned}
x(P + P_0) &= \left(\frac{y_0 - y}{x_0 - x}\right)^2 - x - x_0 \\
&= \frac{y_0^2 - 2y_0 y + y^2}{(x_0 - x)^2} - \frac{(x_0 + x)}{(x_0 - x)^2}(x_0^2 - 2x_0 x + x^2) \\
&= \frac{x_0^3 + ax_0 + b + x^3 + ax + b - 2y_0 y}{(x_0 - x)^2} \\
&\quad - \frac{(x^3 - 2x_0^2 x + x_0 x^2 + x_0 x^2 + xx_0^2 - 2x_0 x^2 + x^3)}{(x_0 - x)^2} \\
&= \frac{a(x_0 + x) + 2b - 2y_0 y - (-x_0^2 x - x^2 x_0)}{(x_0 - x)^2} \\
&= \frac{a(x_0 + x) + 2b - 2y_0 y + x_0 x(x_0 + x)}{(x_0 - x)^2} \\
&= \frac{(a + x_0 x)(x_0 + x) + 2b - 2y_0 y}{(x_0 - x)^2}.
\end{aligned}
$$

Substituting r, s, t then gives

$$x(P + P_0) = \frac{\left(a + \frac{r_0 r}{t_0^2 t^2}\right)\left(\frac{r_0}{t_0^2} + \frac{r}{t^2}\right) + 2b - 2\frac{ss_0}{t_0^3 t^3}}{\left(\frac{r_0}{t_0^2} - \frac{r}{t^2}\right)^2}$$

$$= \frac{(at_0^2 t^2 + r_0 r)(r_0 t^2 + r t_0^2) + 2b t^4 t_0^4 - 2ss_0 t t_0}{(r_0 t^2 - r t_0^2)^2}.$$

The effect of clearing denominators in the rationals a and b appearing as coefficients in the elliptic curve can be absorbed into the constant c_1.

It is therefore sufficient to check that the numerator and denominator satisfy the inequality in the lemma.

$$|\text{numerator}| < \underbrace{(c_2|t|^2 + c_3|r|)(c_4|t|^2 + c_5|r|)}_{<c_8(\max\{|r|,|t|^2\})^2} + c_6|t|^4 + c_7|st|. \qquad (5.10)$$

The first two terms have

$$c_8(\max\{|r|, |t|^2\})^2 + c_6|t|^4 \le c_9 H(P)^2.$$

Looking at the third term, we need to show that

$$|st| \le c_{10} H(P)^2.$$

Since $y^2 = x^3 + ax + b$ we have

$$(st)^2 = r^3 t^2 + art^6 + bt^8.$$

There are two cases to consider.

Case I: $|r| \ge |t|^2$. In this case

$$|st|^2 \le c_{11}|r|^4 + c_{12}|r|^4 + c_{13}|r|^4 = c_{14} H(P)^4.$$

Case II: $|r| < |t|^2$. In this case

$$|st|^2 < c_{15}|t|^8 + c_{16}|t|^8 + c_{17}|t|^8 = c_{18} H(P)^4.$$

In both cases $|st| < c_{10} H(P)^2$ as required. Therefore

$$|\text{numerator}| < c_{19} H(P)^2,$$

or in logarithmic form

$$\log|\text{numerator}| < c_{20} + 2h(P).$$

The denominator is simpler:

$$|\text{denominator}| = |r_0 t^2 - r t_0^2|^2$$
$$< \left(c_{21}|t|^2 + c_{22}|r|\right)^2$$
$$< c_{23} H(P)^2,$$

so

$$\log|\text{denominator}| < c_{24} + 2h(P).$$

We will assume the *weak Mordell theorem*. This has a lengthy proof that would take us too far away from the basic idea of heights.

Theorem 5.20. [WEAK MORDELL THEOREM] *Let E denote an elliptic curve defined over $\mathbb{Q}$. Then $E(\mathbb{Q})/2E(\mathbb{Q})$ is a finite abelian group.*

Proof. See Lang [Lan78] or Silverman [Sil86]. An outline proof from an advanced point of view is in Milne [Mil86, Theorem 20.10].

Proof (of Theorem 5.12 assuming Theorem 5.20). Let $\mathcal{Q} = \{Q_1, \ldots, Q_s\}$ denote a fixed set of coset representatives for $2E(\mathbb{Q})$ in $E(\mathbb{Q})$. It is enough to prove the following: there is a constant $R = R(E, \mathcal{Q})$ with the property that every point $P \in E(\mathbb{Q})$ can be expressed as an integral combination of the Q_i, $i = 1, \ldots, s$ and those $Q \in E(\mathbb{Q})$ with $h(Q) < R$ (finite in number by Lemma 5.19[2]).

Let P be a rational point on E, and write (for some $i_0, i_1, \ldots \in \{1, \ldots, s\}$)

$$P = P_0 = Q_{i_0} + 2P_1$$
$$P_1 = Q_{i_1} + 2P_2$$
$$\vdots$$
$$P_n = Q_{i_n} + 2P_{n+1}.$$

The duplication formula (5.6) (Theorem 5.13) says that

$$4h(P_{n+1}) - c_1 < h(2P_{n+1}) \tag{5.11}$$

for some $c_1 = c_1(E) > 0$. On the other hand, Lemma 5.19[1] shows that

$$h(2P_{n+1}) = h(P_n - Q_{i_n}) < 2h(P_n) + c_2 \tag{5.12}$$

for some $c_2 = c_2(E, \mathcal{Q})$. Combining (5.11) and (5.12) gives

$$h(P_{n+1}) < \tfrac{1}{2}h(P_n) + c_3$$

for some $c_3 = c_3(E, \mathcal{Q})$. Iterating this gives

$$h(P_{n+1}) < \tfrac{1}{2}\left(\tfrac{1}{2}h(P_{n-1}) + c_3\right) + c_3$$
$$= \tfrac{1}{2^2}h(P_{n-1}) + c_3\left(1 + \tfrac{1}{2}\right)$$
$$\vdots$$
$$< \tfrac{1}{2^{n+1}}h(P_0) + c_3\left(1 + \tfrac{1}{2} + \tfrac{1}{2^2} + \ldots + \tfrac{1}{2^n}\right).$$

As $n \to \infty$, for fixed P_0

$$\frac{1}{2^{n+1}}h(P_0) \to 0 \text{ and } c_3\left(1 + \tfrac{1}{2} + \tfrac{1}{2^2} + \ldots + \tfrac{1}{2^n}\right) \to 2c_3. \tag{5.13}$$

Now $P = Q_{i_0} + 2P_1$, $P_1 = Q_{i_1} + 2P_2$ and so on gives

$$P = Q_{i_0} + 2(Q_{i_1} + 2P_2)$$
$$= Q_{i_0} + 2Q_{i_1} + 2^2 P_2$$
$$= Q_{i_0} + 2Q_{i_1} + 2^2 Q_{i_2} + 2^3 P_3$$
$$\vdots$$

so P is being written as an integer combination of elements of Q and a point whose height is uniformly bounded as $n \to \infty$ by (5.13). Take $R = (2 + \frac{1}{10})c_3$, a constant depending on E and on Q. We have shown that any rational point P can be written as an integral combination of the elements of Q and a point with height bounded by R.

Remark 5.21. This proof is very inefficient. It is an open problem to find a minimal generating set for $E(\mathbb{Q})$ given E.

5.7 The Parallelogram Law and the Canonical Height

The duplication formula (5.6) (Theorem 5.13) says that for any $P \in E(\mathbb{Q})$

$$h(2P) = 4h(P) + O(1),$$

or, equivalently, there is a constant $c = c(E)$ such that

$$\left| h(P) - \tfrac{1}{4}h(2P) \right| < c. \tag{5.14}$$

The next result exploits this to produce a height function with better functorial properties, the *canonical height*. The approach below is due to Tate; the canonical height was discovered independently by Neron.

Theorem 5.22. *For any rational point P on an elliptic curve E defined over the rationals,*

$$\lim_{n \to \infty} \frac{h(2^n P)}{4^n} = \hat{h}(P) \tag{5.15}$$

exists. The limit $\hat{h}(P)$ is called the canonical height of P.

Proof. Let $a_N = \frac{1}{4^N} h(2^N P)$. If $N > M \geq 1$, then

$$a_M - a_N = \tfrac{1}{4^M} h(2^M P) - \tfrac{1}{4^N} h(2^N P)$$
$$= \tfrac{1}{4^M} h(2^M P) - \tfrac{1}{4^{M+1}} h(2^{M+1} P)$$
$$\quad + \tfrac{1}{4^{M+1}} h(2^{M+1} P) - \tfrac{1}{4^{M+2}} h(2^{M+2} P)$$
$$\quad + \ldots$$
$$\quad + \tfrac{1}{4^{N-1}} h(2^{N-1} P) - \tfrac{1}{4^N} h(2^N P)$$

which may be grouped into

$$a_M - a_N = \tfrac{1}{4^M} \left(h(2^M P) - \tfrac{1}{4} h(2 \cdot 2^M P) \right)$$
$$\tfrac{1}{4^{M+1}} \left(h(2^{M+1} P) - \tfrac{1}{4} h(2 \cdot 2^{M+1} P) \right)$$
$$+ \ldots$$
$$+ \tfrac{1}{4^{N-1}} \left(h(2^{N-1} P) - \tfrac{1}{4} h(2 \cdot 2^{N-1} P) \right).$$

By the duplication formula (5.14), this gives

$$|a_M - a_N| < \tfrac{1}{4^M} c \left(1 + \tfrac{1}{4} + \tfrac{1}{4^2} + \ldots \right) = \tfrac{1}{4^M} c \left(\tfrac{4}{3} \right) \to 0 \text{ as } M \to \infty,$$

showing that (a_N) is Cauchy.

Remark 5.23. Notice that if the order of P is a power of 2 (see Exercise 5.7 for example) then it is clear that $\hat{h}(P) = 0$. We will prove later that in fact all torsion points P have $\hat{h}(P) = 0$, and conversely that $\hat{h}(P) = 0$ implies that P is a torsion point (see Theorem 5.24[4]).

Theorem 5.24. *Let E be an elliptic curve defined over the rationals.*
[1] For every point $P \in E(\mathbb{Q})$,

$$\hat{h}(P) = h(P) + O(1)$$

uniformly.
[2] [PARALLELOGRAM LAW] For all $P, Q \in E(\mathbb{Q})$,

$$\hat{h}(P + Q) + \hat{h}(P - Q) = 2\hat{h}(P) + 2\hat{h}(Q). \tag{5.16}$$

[3] For every $m \in \mathbb{Z}$ and $P \in E(\mathbb{Q})$

$$\hat{h}(mP) = m^2 \hat{h}(P).$$

[4] For $P \in E(\mathbb{Q})$,

$$\hat{h}(P) = 0 \text{ if and only if } P \text{ is a torsion point.}$$

To prove this a generalization of the duplication formula is needed. One possible generalization would be the formula

$$h(mP) = m^2 h(P) + O(1),$$

but a more useful generalization is the naïve parallelogram law.

Lemma 5.25. *For all $P, Q \in E(\mathbb{Q})$,*

$$h(P + Q) + h(P - Q) = 2h(P) + 2h(Q) + O(1) \tag{5.17}$$

uniformly.

Proof (of Theorem 5.24). [1] By iterating

$$h(P) = \tfrac{1}{4}\left(h(2P) + O(1)\right)$$

we have

$$h(P) = \tfrac{1}{4}\left(\left(\tfrac{h(2^2 P)}{4} + \tfrac{1}{4}O(1)\right) + O(1)\right)$$

$$= \frac{h(2^2 P)}{4^2} + O(1)\left(\tfrac{1}{4} + \tfrac{1}{4^2}\right)$$

$$\vdots$$

$$= \frac{h(2^N P)}{4^N} + \underbrace{O(1)\left(\tfrac{1}{4} + \tfrac{1}{4^2} + \ldots + \tfrac{1}{4^N}\right)}_{O(1)}.$$

Letting $N \to \infty$ gives

$$h(P) = \hat{h}(P) + O(1).$$

[2] Applying a similar limiting procedure to the naïve parallelogram law (5.17) gives

$$\hat{h}(P+Q) + \hat{h}(P-Q) - 2\hat{h}(P) - 2\hat{h}(Q)$$
$$= \lim_{N \to \infty}\left(\tfrac{1}{4^N}h(2^N(P+Q)) + \tfrac{1}{4^N}h(2^N(P-Q))\right.$$
$$\left. - \tfrac{2}{4^N}h(2^N P) - \tfrac{2}{4^N}h(2^N Q)\right)$$
$$= \lim_{N \to \infty}\left(\tfrac{1}{4^N}O(1)\right) = 0.$$

[3] This is proved by induction on $m \geq 1$ (the case $m \leq -1$ follows since $h(-P) = h(P) \Rightarrow \hat{h}(P) = \hat{h}(-P)$). For $m = 0$ $h(0) = 0 = \hat{h}(0)$. Assume therefore that

$$\hat{h}(mP) = m^2 \hat{h}(P),$$

and substitute mP for P, P for Q in the parallelogram law (5.16):

$$\hat{h}(mP + P) = 2\hat{h}(mP) + 2\hat{h}(P) - \hat{h}((m-1)P)$$
$$= 2m^2\hat{h}(P) + 2\hat{h}(P) - (m-1)^2\hat{h}(P)$$
$$= (m+1)^2\hat{h}(P).$$

[4] If P is a torsion point then $mP = 0$ for some $m \neq 0$ so by [3] $\hat{h}(P) = 0$. Conversely, suppose that $\hat{h}(P) = 0$ for some $P \in E(\mathbb{Q})$. Then $\hat{h}(mP) = m^2\hat{h}(P) = 0$ for all m, so $h(mP)$ must be uniformly bounded for all m by [1]. By Lemma 5.19[2], this means that the set $\{mP\}_{m \in \mathbb{Z}}$ must be finite, so P is a torsion point.

Proof (of Lemma 5.25). Let $E : y^2 = x^3 + ax + b$ be the elliptic curve. Let $P, Q \in E(\mathbb{Q})$, and write $x(P) = x_1, x(Q) = x_2, x(P+Q) = x_3$, and $x(P - Q) = x_4$.

The values of x_3 and x_4 depend on the y coordinates of P and Q; this complicates the proof. We will work in the coordinates $x_1 x_2$, $x_1 + x_2$, $x_3 x_4$ and $x_3 + x_4$ because these only depend on the x coordinates. Now

$$x_3 + x_4 = \frac{2(x_1 + x_2)(a + x_1 x_2) + 4b}{(x_1 - x_2)^2},$$

$$x_3 x_4 = \frac{(x_1 x_2 - a)^2 - 4b(x_1 + x_2)}{(x_1 - x_2)^2}$$

and we may write

$$(x_1 - x_2)^2 = (x_1 + x_2)^2 - 4x_1 x_2$$

giving $x_3 + x_4$ and $x_3 x_4$ in terms of $x_1 x_2$, $x_1 + x_2$.

Now we claim that for any $x_1, x_2 \in \mathbb{Q}$,

$$h([x_1 + x_2, x_1 x_2, 1]) = h([x_1, 1]) + h([x_2, 1]) + O(1). \tag{5.18}$$

To see this, write $x_1 = \frac{s}{t}, x_2 = \frac{u}{v}$ in lowest terms, and define

$$F_1(x) = tx - s, \quad F_2(x) = vx - u.$$

Then

$$m(F_1 F_2) = m(F_1) m(F_2). \tag{5.19}$$

Now by Mahler's result (equation (1.6) in Remark 1.5, and Exercise 1.3),

$$m\left(\sum_{i=0}^{d} a_i x^i\right) = h\left(\sum_{i=0}^{d} a_i x^i\right) + O(d) \tag{5.20}$$

where $h(\sum_{i=0}^{d} a_i x^i) = \log \max\{|a_i|\}$. Applying this to (5.19) gives

$$h(F_1 F_2) = h(F_1) + h(F_2) + O(1). \tag{5.21}$$

Now

$$h(F_1) = h([x_1, 1]), \quad h(F_2) = h([x_2, 1]). \tag{5.22}$$

On the other hand

$$F_1(x) F_2(x) = (tx - s)(vx - u) = tvx^2 - x(sv + tu) + su$$

so

$$h(F_1 F_2) = \max\{|tv|, |sv + tu|, |su|\}.$$

Now

$$x_1 + x_2 = \frac{sv + tu}{tv}$$

$$x_1 x_2 = \frac{su}{tv}$$

so

$$h([x_1 + x_2, x_1x_2, 1]) = h([\frac{sv + tu}{tv}, \frac{su}{tv}, 1]) = h([sv + tu, su, tv]).$$

Now $sv + tu, su$ and tv cannot have a common factor by Gauss' lemma, so $h(F_1 F_2) = h([x_1 + x_2, x_1x_2, 1])$, and (5.21), (5.22) gives (5.18).

So we can change variables and work with $x_1 + x_2$ and $x_1 x_2$:

$$x_3 + x_4 = \frac{2(x_1 + x_2)(a + x_1x_2) + 4b}{(x_1 + x_2)^2 - 4x_1x_2}$$
$$x_3 x_4 = \frac{(x_1 x_2 - a)^2 - 4b(x_1 + x_2)}{(x_1 + x_2)^2 - 4x_1x_2}.$$

Lemma 5.26. *Assume that $4a^3 + 27b^2 \neq 0$. Then the map $\mathbb{P}^2(\mathbb{Q}) \to \mathbb{P}^2(\mathbb{Q})$ defined by*

$$[u, v, t] = [2u(at + v) + 4bt^2, (v - at)^2 - 4btu, u^2 - 4tv]$$

is a morphism of degree 2.

The formulæ in Lemma 5.26 come from setting $u = x_1 + x_2$, $v = x_1 x_2$ and using t to make the expressions homogeneous.

Proof (of Lemma 5.26). Suppose the three polynomials vanish:

$$2u(at + v) + 4bt^2 = 0$$
$$(v - at)^2 - 4btu = 0$$
$$u^2 - 4tv = 0. \tag{5.23}$$

If $t = 0$ then $u = 0$ by (5.23) so $v = 0$ by (5.23). Suppose therefore that $t \neq 0$, and divide by t^2 in each equation. Write $x = \frac{u}{2t}$, so $x^2 = \frac{v}{t}$ by (5.23). Equations (5.23) and (5.23) give

$$(x^2 - a)^2 - 8bx = 0$$
$$x(a + x^2) + b = 0$$

or

$$x^4 - 2ax^2 - 8bx + a^2 = 0$$
$$x^3 + ax + b = 0.$$

These polynomials arose in the proof of the duplication formula (cf. equation (5.9)), where it was shown that they have no common zero.

Now apply Lemma 5.26 to the vectors

$$[x_1 + x_2, x_1x_2, 1] \text{ and } [x_3 + x_4, x_3x_4, 1].$$

Since the map from the first to the second is a morphism of degree 2,

$$h([x_3 + x_4, x_3 x_4, 1]) = 2h([x_1 + x_2, x_1 x_2, 1]) + O(1)$$

by Lemma 5.16. Equation (5.18) shows that

$$h([x_3 + x_4, x_3 x_4, 1]) = h([x_3, 1]) + h([x_4, 1]) + O(1)$$

and

$$h([x_1 + x_2, x_1 x_2, 1]) = h([x_1, 1]) + h([x_2, 1]) + O(1),$$

so

$$h([x_3, 1]) + h([x_4, 1]) = 2h([x_1, 1]) + 2h([x_2, 1]) + O(1),$$

and therefore

$$h(P + Q) + h(P - Q) = 2h(P) + 2h(Q) + O(1).$$

5.8 Heights of Algebraic Numbers

In arithmetic geometry, it is more common to meet the notion of height attached to an algebraic number or to an algebraic point on a variety. There is a very good reason for this, which goes back to Weil's generalization of Mordell's Theorem 5.12 in [Wei28], [Wei30]. For any number field K, the group of K-rational points on an elliptic curve defined over K forms a finitely generated group. In order to prove this, Weil had to define a notion of height which generalized Mordell's. This requires some imagination, and we indicate here how it is done.

In addition to the usual valuation $x \mapsto |x|$ on $\mathbb{Q}$, there are p-adic valuations $x \mapsto |x|_p$, one for each prime number p. If p is a prime number, then define

$$|\alpha|_p \overset{.}{=} p^{-\operatorname{ord}_p(\alpha)},$$

where $\operatorname{ord}_p(\alpha)$ is the (signed) power of p that divides α. The definition is extended to all of $\mathbb{Q}$ by defining $|0|_p = 0$. Valuations are sometimes called 'absolute values', and we will do that when appropriate here.

Example 5.27. We have $\operatorname{ord}_3(5) = 0$, $\operatorname{ord}_7(3/7) = -1$, $\operatorname{ord}_3(3/7) = 1$, and $\operatorname{ord}_5(1/25) = -2$, so $|5|_3 = 1$, $|3/7|_7 = 7$, $|3/7|_3 = 1/3$, and $|1/25|_5 = 25$ respectively.

Exercise 5.15. Given a prime p, verify that $|\cdot|_p$ satisfies the axioms for an absolute value:
[1] $|\alpha|_p = 0$ if and only if $\alpha = 0$;
[2] $|\alpha\beta|_p = |\alpha|_p |\beta|_p$;
[3] $|\alpha + \beta|_p \leq |\alpha|_p + |\beta|_p$.
    Show that the triangle inequality [3] may be replaced by the stronger ultrametric inequality
[3*] $|\alpha + \beta|_p \leq \max\{|\alpha|_p, |\beta|_p\}$.
    Which properties fail if the same definition is made for p non-prime?

Theorem 5.28. [PRODUCT FORMULA] *For all* $\alpha \in \mathbb{Q} \backslash \{0\}$

$$|\alpha| \cdot \prod_p |\alpha|_p = 1. \tag{5.24}$$

Exercise 5.16. Prove the product formula.

It is sometimes convenient to put all the absolute values on the same footing, associating the usual absolute value to an 'infinite prime', and writing $|\cdot| = |\cdot|_\infty$. The product formula (5.24) may then be written

$$\prod_{p \le \infty} |\alpha|_p = 1 \text{ for } \alpha \in \mathbb{Q} \backslash \{0\}.$$

The usual absolute value is often referred to as 'archimedean' since it satisfies [3] but not [3*] in Exercise 5.15 above. The inequality [3] is usually called the 'triangle inequality' since for the complex numbers and the usual absolute value it says that in any triangle the sum of the lengths of two sides is greater than or equal to the length of the third. Property [3*], which holds for all the p-adic absolute values, says that all triangles are isosceles in some sense (see Koblitz's book [Kob77, Chapter I.2] for an explanation of this). The p-adic absolute values are said to be 'non-archimedean'.

Definition 5.29. For any point $P = [x_0, \dots, x_N] \in \mathbb{P}^N(\mathbb{Q})$, define

$$H(P) = \prod_{p \le \infty} \max\{|x_0|_p, \dots, |x_N|_p\}; \quad h(P) = \log H(P).$$

Notice that choosing a representative for P in which the x_i are coprime integers, we recover the definition (5.8). However, by the product formula, the formula in Definition 5.29 is invariant across all representatives of a given point.

It is this set-up which is often taken as the starting point for generalization. Given an algebraic number field K, all the valuations on $\mathbb{Q}$ can be extended to K and these can be used to define a notion of height for elements of $\mathbb{P}^N(K)$. Our point of view has been different because both pedagogically and mathematically it is polynomials which arise first. There are also technical simplifications in working with heights of polynomials. To indicate some of the inter-relations between the different notions and what has been deliberately suppressed, consider the rational number $\alpha = b/a$ (in lowest terms).

We may associate to α the integer polynomial $F(x) = ax - b$, with Mahler measure $m(F) = \log \max\{|a|, |b|\}$. On the other hand, we may associate to α the point $P = [1, b/a] \in \mathbb{P}^1(\mathbb{Q})$, whose height $h(P)$ in the sense of Definition 5.29 is again $m(F)$. Notice that all the non-archimedean data about P (or F) is contained in the integer a. This was used at several points in Chapters 1 and 2, and in a more subtle way in Chapters 3 and 4. For example, in Zagier's proof of Zhang's theorem (Section 1.5), rather than argue with heights as he does, we have used the archimedean absolute value and then added $\log|a|$ to both sides. In the proof of Theorem 2.12 all the non-archimedean behaviour is again grouped together into a single number.

Arithmetic which takes account of all the different valuations simultaneously is *adelic*. This is a very powerful point of view, and in several places in this account, the adelic point of view can illuminate and simplify (at some technical cost). Theorem 2.12 has a proof entirely from this viewpoint, due to Lind and Ward [LW88] and this gives a very complete picture of the local hyperbolicity along the lines of the special case discussed after Theorem 2.6. Once again all the non-archimedean contributions (to the entropy in this case) group together into one simple '$\log|a|$' term.

In a similar spirit, the canonical height is composed of local heights, one for each valuation, and the denominator of a rational point is essentially the non-archimedean part of the height (cf. proof of Theorem 6.6). Of course the expected 'elliptic' dynamical systems should behave adelically (cf. [EW98]).

Returning to algebraic numbers, if $\alpha \in \bar{\mathbb{Q}}$ is non-zero then it has a minimal polynomial $F \in \mathbb{Z}[x]$ of degree d say. It turns out that $h(\alpha) = \frac{1}{d}m(F)$, so the concept of height of an algebraic number and the measure of a polynomial are intimately related. This makes the elliptic analogue of Lehmer's problem a natural question: if E is an elliptic curve defined over $\mathbb{Q}$ and $P \in E(\bar{\mathbb{Q}})$ is a non-torsion point with degree of x-coordinate d, is $\hat{h}(P) > \frac{C}{d}$ for some constant $C > 0$ depending only on E? The elliptic analogue of Dobrowolski's bound was proved by Laurent [Lau83] in a special case and by Masser [Mas89] in general (with a weaker estimate) but the full question seems even less accessible than the original question. See [Eve] for a discussion of the elliptic Lehmer problem in the elliptic non-reciprocal case and Question 11 in Chapter 6 for a strengthened elliptic bound in the totally real case (cf. Theorem 1.14).

6. The Elliptic Mahler Measure

This chapter has three themes. In Section 6.1 a very short introduction to the classical theory of elliptic functions is given. In Sections 6.2, 6.3 and 6.5 some recent work on the elliptic Mahler measure is presented. Sections 6.4 and 6.6 contain evidence for a possible family of dynamical systems associated to elliptic curves whose dynamical properties are linked to the elliptic Mahler measure in a manner analogous to the connection between Chapters 1 and 3 and Chapters 2 and 4. This third theme is more speculative in nature.

6.1 Elliptic Functions

Underlying Mahler's measure is a compact abelian group, namely, the circle. This group can be represented as a complex curve and is parametrized by the values of a transcendental function, the exponential. In this chapter this group is replaced by another well-known compact abelian group, the group of complex points of an elliptic curve. This group is also parametrized by the values of a transcendental function, known as an *elliptic function*.

Let $L \subset \mathbb{C}$ denote a lattice in the complex numbers. This means L is the set of integer linear combinations of two complex numbers w_1 and w_2 which are linearly independent over $\mathbb{R}$. The lattice L plays the role of the set of integer multiples of 2π in $\mathbb{R}$. The exponential function $x \mapsto e^{ix}$ satisfies $e^{i(x+2\pi m)} = e^{ix}$ for all $x \in \mathbb{R}, m \in \mathbb{Z}$ so the exponential is periodic with respect to the one-dimensional lattice $2\pi\mathbb{Z}$. We are interested in complex functions which are periodic with respect to L.

The lattice L is represented as a discrete subset of $\mathbb{C}$ in Figure 6.1. The shaded region $\Pi = \{r_1 w_1 + r_2 w_2 : 0 \le r_1, r_2 < 1\}$ is known as a *fundamental domain*. The basic analogue of the exponential function is the Weierstrass $\wp$-function corresponding to L. For $z \notin L$, this is defined by the series

$$\wp_L(z) = \frac{1}{z^2} + \sum_{0 \neq \ell \in L} \left\{ \frac{1}{(z-\ell)^2} - \frac{1}{\ell^2} \right\}. \tag{6.1}$$

We are going to show that $\wp_L(z)$ is periodic with respect to L. That is

$$\wp_L(z + \ell) = \wp_L(z) \text{ for all } \ell \in L.$$

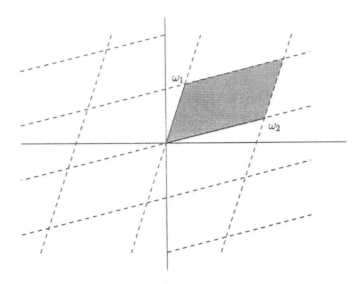

Fig. 6.1. The lattice $L \subset \mathbb{C}$

This property is not obvious: proving it from the definition (6.1) involves checking that the series converges absolutely in order to justify rearrangements of the terms.

Lemma 6.1. *The series defining $\wp_L(z)$ is absolutely convergent for all z not in L. The series defines a meromorphic function whose only singularities are double poles at each lattice point $\ell \in L$.*

Proof. Given any $z \in \mathbb{C} \backslash L$, write

$$\frac{1}{(z-\ell)^2} - \frac{1}{\ell^2} = \frac{1}{\ell^3} \cdot \frac{2z - z^2/\ell}{(z/\ell - 1)^2}.$$

Since $z \notin L$, it follows that $|z/\ell| \to 0$ as $|\ell| \to \infty$. Therefore, for some $C_1 = C_1(z)$,

$$\left| \frac{1}{(z-\ell)^2} - \frac{1}{\ell^2} \right| \le \frac{C_1}{|\ell|^3}.$$

Therefore, it is sufficient to prove that the series $\sum_{0 \ne \ell \in L} |\ell|^{-3}$ converges. Write the series as

$$\sum_{n=1}^{\infty} \sum_{\substack{\ell \in L \\ n \le |\ell| < n+1}} |\ell|^{-3}.$$

Note that we have omitted those ℓ with $|\ell| < 1$. This is a finite set so the removal of it does not affect the convergence of the series. If $n \le |\ell|$ then $1/|\ell|^3 \le 1/n^3$ so the sum is bounded by

$$\sum_{n=1}^{\infty} |\{\ell \in L : n \le |\ell| < n+1\}|/n^3. \qquad (6.2)$$

The number of $\ell \in L$ with $n \le |\ell| < n+1$ can be estimated using a geometric argument. They lie in the annulus between the two discs of radii n and $n+1$. The number of elements $\ell \in L$ with $|\ell| < R$ is clearly

$$\frac{\pi R^2}{A} + O(R),$$

where A is the area of the fundamental domain Π. Applying this formula to our two circles and subtracting the result gives an upper bound of $C_2 n$ for those $\ell \in L$ with $n \le |\ell| < n+1$. Thus the sum in (6.2) is bounded above by $C_2 \sum_{n=1}^{\infty} n^{-2}$ which is known to converge.

This shows that $\wp_L(z)$ converges absolutely for $z \in \mathbb{C}\backslash L$. A similar argument, applied to $\wp_L(z) - 1/(z-\ell)^2$, shows that the only poles of $\wp_L(z)$ are double poles at each of the lattice points.

The absolute convergence of $\wp_L(z)$ means that (6.1) can be differentiated term-by-term to obtain

$$\wp'_L(z) = -2 \sum_{\ell \in L} \frac{1}{(z-\ell)^3},$$

which also converges absolutely. Clearly $\wp'_L(z)$ is periodic since we can rearrange the terms. We can now exploit the periodicity of $\wp'_L(z)$ to deduce the periodicity of $\wp_L(z)$.

Lemma 6.2. *The Weierstrass $\wp$-function $\wp_L(z)$ is periodic with respect to L.*

Proof. It is enough to prove that $\wp_L(z + w_i) = \wp_L(z)$ for $i = 1, 2$. Note first that $\wp_L(-z) = \wp_L(z)$ and $\wp'_L(-z) = -\wp'_L(z)$. (In other words, $\wp_L(z)$ is *even* and $\wp'_L(z)$ is *odd*.) To prove the periodicity, let f denote the function

$$f(z) = \wp_L(z + w_i) - \wp_L(z).$$

Then f is differentiable for all $z \notin L$. Using the periodicity of $\wp'_L(z)$, we deduce that $f'(z) = 0$ for all $z \in \mathbb{C}\backslash L$. Thus f is constant on this open set. To determine that constant let $z = -w_i/2$. Then

$$f(-w_i/2) = \wp_L(w_i/2) - \wp_L(-w_i/2).$$

Using the evenness of $\wp_L(z)$, we see that $f(w_i/2) = 0$. Since f is a constant function, it must be zero everywhere.

What this result says is that $\wp_L(z)$ and $\wp_L'(z)$ are *elliptic functions* (for L). This is the name given to meromorphic functions which are periodic with respect to L. We will now record some simple facts about elliptic functions.

Lemma 6.3. [1] *An elliptic function with no poles in Π is constant.*
[2] *Let $\Pi_\beta = \beta + \Pi$ be the displaced parallelogram, and let f denote an elliptic function with no zeros or poles on the boundary of Π_β. If the zeros of f in Π_β have orders m_i and the poles have orders n_j then $\sum m_i = \sum n_j$.*

Proof. Part [1] is obvious: any such function would be bounded on Π, hence on $\mathbb{C}$ by periodicity. But a bounded entire function is constant.

For part [2], note first that $\int_{\Pi_\beta} f(z)dz = 0$ since f has the same values on opposite sides of Π_β while dz changes sign. The result now follows by applying this to f'/f. Near a zero of order m, this function looks like $\frac{m}{z-z_0}$; near a pole of order n it looks like $-\frac{n}{z-z_0}$. Cauchy's residue theorem gives the result.

We will use Lemma 6.3 to prove the main result, that the values of $\wp_L(z)$ and $\wp_L'(z)$, for z lying in the fundamental domain, parametrize a complex curve.

Theorem 6.4. [1] *There are constants $a = a(L)$ and $b = b(L)$ with $4a^3 + 27b^2 \neq 0$, such that for all $z \in \mathbb{C} \backslash L$,*

$$\tfrac{1}{4}\wp_L'(z)^2 = \wp_L(z)^3 + a\wp_L(z) + b.$$

[2] *For $z \in \mathbb{C}/L$, the map $\pi : \Pi \to \mathbb{P}^2(\mathbb{Q})$ defined by $\pi(0) = [0, 1, 0]$ and*

$$\pi(z) = [\wp_L(z), \tfrac{1}{2}\wp_L'(z), 1], \quad z \neq 0,$$

defines a bijection between Π and the set of complex projective points on the elliptic curve $E : y^2 = x^3 + ax + b$.
[3] *Suppose $z_1, z_2, z_3 \in \Pi$ have images $\pi(z_i) = P_i, i = 1, 2, 3$. Then $z_1 + z_2 + z_3 = 0$ in Π if and only if P_1, P_2 and P_3 lie on a straight line.*

Write $\langle \omega_1, \omega_2 \rangle$ for the subgroup $\omega_1 \mathbb{Z} + \omega_2 \mathbb{Z} \subset \mathbb{C}$.

Exercise 6.1. [1] Let $L = \langle 1, i \rangle$. Then the corresponding elliptic curve E_L has equation $y^2 = x^3 + ax$ for some $a \in \mathbb{R}$.
[2] Let $L = \langle 1, \omega \rangle$, where ω denotes a cube root of unity. The corresponding elliptic curve E_L has equation $y^2 = x^3 + b$ for some $b \in \mathbb{R}$.

Under the bijection in [2], the fact that the point $0 = z \in \mathbb{C}/L$ corresponds to the point at infinity marries neatly the geometric idea of infinity on the projective curve to the analytic idea that $\wp_L(z) \to \infty$ as $z \to 0$. This is important if we work with the projective curve because the set of projective points form a group with the point at infinity as the identity. This arises simply by transporting the group structure of $\mathbb{C}/L$ to the curve. The marvel of this is part [3]. That is, the simple addition in $\mathbb{C}$ is related, via some transcendental functions, to the geometric addition on the projective curve. Note that it also takes account of multiplicities. This transport of structure proves that the geometric group law on the projective curve really does satisfy the axioms for a group. If you allow the 'Lefschetz principle' then it also verifies the group law for elliptic curves over arbitrary fields in characteristic not equal to 2 or 3 (the Lefschetz principle is discussed in Silverman [Sil86, Section VI.6]).

An important application of Theorem 6.4 is that it allows the torsion points on an elliptic curve to be understood in a way which is analogous to our understanding of torsion points on the circle. Given $1 \leq n \in \mathbb{N}$, the points $z = (r_1\omega_1 + r_2\omega_2)/n$ for $0 \leq r_1, r_2 \leq n$ all have $nz \equiv 0 \mod L$, and these are the n^2 points with order dividing n. Of course these are torsion points on the *complex* curve: which of these points correspond to rational points on the curve is a different question. Thus, Section 5.3 may be interpreted as follows: for an elliptic curve E defined over $\mathbb{Q}$, of the 121 points P with $11P = 0$ in $E(\mathbb{C})$, only the identity is also in $E(\mathbb{Q})$.

Exercise 6.2. Let $E(\mathbb{C})$ denote the complex points on a complex elliptic curve and let $E_n(\mathbb{C})$ for $n \in \mathbb{N}$ denote the subgroup of torsion points whose order divides n. Prove that

$$E_n(\mathbb{C}) \cong \mathbb{Z}/n\mathbb{Z} \oplus \mathbb{Z}/n\mathbb{Z}.$$

Exercise 6.3. Show that

$$\wp'_L(\omega_1/2) = \wp'_L(\omega_2/2) = \wp'_L((\omega_1 + \omega_2)/2) = 0.$$

We are going to present a proof to show how the equation in Theorem 6.4[1] comes about. We will not prove that the discriminant $4a^3 + 27b^2$ is nonzero. Neither will we prove that a kind of converse is true: namely, that if a and b are given with $4a^3 + 27b^2 \neq 0$ then there exists a lattice L such that $\wp_L(z)$ and $\frac{1}{2}\wp'_L(z)$ Parametrize the elliptic curve with equation $y^2 = x^3 + ax + b$ (see Koblitz [Kob84] or Silverman [Sil86] and [Sil94] for a full account).

Proof (of Theorem 6.4).
   [1] Go back to the expression $\frac{1}{(z-\ell)^2} - \frac{1}{\ell^2}$. If $|z| < |\ell|$ for all non-zero ℓ then we can expand this expression as a Taylor series about $z = 0$:

$$\frac{1}{(z-\ell)^2} - \frac{1}{\ell^2} = \frac{1}{\ell^2}(1 - z/\ell)^{-2} = \frac{2z}{\ell^3} + \frac{3z^2}{\ell^4} + \frac{4z^3}{\ell^5} + \dots$$

Using the absolute convergence of the series defining $\wp_L(z)$, we can rearrange the terms in

$$\wp_L(z) = \frac{1}{z^2} + \sum_{0 \neq \ell \in L} \left(\frac{2z}{\ell^3} + \frac{3z^2}{\ell^4} + \frac{4z^3}{\ell^5} + \dots \right)$$

to get

$$\wp_L(z) = \frac{1}{z^2} + 2z\sum{}'\ell^{-3} + 3z^2\sum{}'\ell^{-4} + 4z^3\sum{}'\ell^{-5} + \ldots.$$

The clicks remind us that the sum is over non-zero lattice points $\ell \in L$. Notice that $\sum'\ell^{-2n-1} = 0$, for all $1 \leq n \in \mathbb{N}$. This is because all the terms cancel out as $(-\ell)^{-2n-1} = -\ell^{-2n-1}$ for all $0 \neq \ell \in L$. In this way, we obtain the Laurent expansion of $\wp_L(z)$ about $z = 0$ in quite an explicit form,

$$\wp_L(z) = \frac{1}{z^2} + 3z^2 G_4(L) + 5z^4 G_6(L) + \ldots, \quad G_{2n}(L) = \sum{}'\ell^{-2n}, 1 \leq n \in \mathbb{N}.$$

Notice how this expression agrees with the classical result that even meromorphic functions have Laurent expansions with only even powers of z.

Now the algebraic differential equation arises quite easily. Consider the function

$$g(z) = \wp_L'(z)^2 - 4\wp_L(z)^3 + 60G_4(L)\wp_L(z) + 140G_6(L).$$

With some careful checking, it will be seen that g has no poles in Π, because the Laurent expansion begins in the z^2 term. Clearly g is periodic so it follows from Lemma 6.3[1] that g must be a constant. Setting $z = 0$ shows that the constant is zero. Thus g is the zero function and we deduce the equation stated in the theorem after dividing by 4. Notice that $a = -15G_4(L)$ and $b = -35G_6(L)$.

[2] There is a little more to this than meets the eye. For example, we know that for $z \notin L$, $\wp_L(-z) = \wp_L(z)$. This does not contradict [2] because $\wp_L'(-z) = -\wp_L'(z)$ so z and $-z$ produce two distinct points on the curve. The only counter-example occurs when $\wp_L'(z) = 0$, but this happens precisely when $z = w_1/2, w_2/2$ or $(w_1 + w_2)/2$ by Exercise 6.3. However, in each of these cases, $2z \in L$ so the points z and $-z$ are not distinct. Also, given any $\alpha \in \mathbb{C}$, the function $\wp_L(z) - \alpha$ has two poles in Π so, by Lemma 6.3[2], it must have two zeros. This does not give two pre-images however by the same argument, providing the zeros are distinct. In the case of coincident roots, we must have $\wp_L'(z) = 0$ so once again we encounter the half-lattice points.

[3] Let the equation of the line containing the points P_1 and P_2 be $y = mz + b$. Consider the function $f(z) = \wp_L'(z) - m\wp_L(z) - b$. This has three poles in Π so, by Lemma 6.3[2], it has three zeros. Two of these are z_1 and z_2. Let z_3 denote the third. Then P_1, P_2 and P_3 lie on the line $y = mz + b$ and part [3] comes about by integrating the function $h(z) = zf'(z)/f(z)$ over a displaced parallelogram $\Pi_\beta = \beta + \Pi$, where β is chosen so that h has no singularities on the boundary Γ_β of Π_β. The heart of the proof is to show that $z_1 + z_2 + z_3 \in L$. To see this, note that, by Cauchy's theorem

$$\frac{1}{2\pi i}\int_{\Gamma_\beta} h(z)dz = z_1 + z_2 + z_3 \tag{6.3}$$

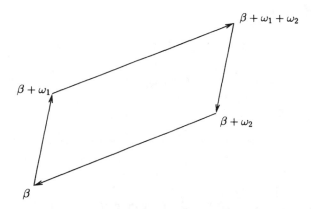

Fig. 6.2. Integrating around Γ_β

because h has a simple pole at each z_i with residue z_i. On the other hand, we can integrate along opposite sides of Γ_β.

For example, take the two sides shown in Figure 6.2 to see that (6.3) expands to

$$\frac{1}{2\pi i}\left(\int_\beta^{\beta+w_1}\frac{zf'(z)}{f(z)}dz+\int_{\beta+w_1+w_2}^{\beta+w_2}\frac{zf'(z)}{f(z)}dz\right).$$

In the second term, change the variable and use the periodicity to obtain

$$\frac{1}{2\pi i}\left(\int_\beta^{\beta+w_1}\frac{zf'(z)}{f(z)}dz-\int_\beta^{\beta+w_1}\frac{(z+w_2)f'(z)}{f(z)}dz\right).$$

After cancellation, this becomes

$$\frac{w_2}{2\pi i}\int_\beta^{\beta+w_1}\frac{f'(z)}{f(z)}dz.$$

Now make the substitution $u=f(z)$. The integral becomes

$$\frac{w_2}{2\pi i}\int_\Omega\frac{1}{u}du,\tag{6.4}$$

where Ω is the image of the line joining β to $\beta+w_1$ in u-space. The periodicity means that Ω is a closed curve, so

$$\frac{1}{2\pi i}\int_\Omega\frac{1}{u}du=m\in\mathbb{Z}.$$

The integer m is the winding number, counting the number of times Ω winds around zero. Thus the integral in (6.4) is mw_2. A similar argument for the other two sides of Π_β shows that contribution is nw_1 for some $n\in\mathbb{Z}$. Thus

$$z_1+z_2+z_3=nw_1+mw_2\in L.$$

Exercise 6.4. [1] Prove that for any lattice $L \subset \mathbb{C}$,

$$G_8(L) = \tfrac{3}{7}G_4^2(L).$$

[2] More generally, prove that all the G_i ($i \geq 8$) can be expressed as polynomials in G_4 and G_8 with rational coefficients. For an explanation of this remarkable phenomenon, consult Koblitz [Kob84].

Exercise 6.5. [1] Given any non-zero $c \in \mathbb{C}$, consider the map $L \to cL = L'$. Let E and E' denote the elliptic curves $\mathbb{C}/L$ and $\mathbb{C}/L'$. Prove that the map produces an isomorphism between elliptic curves E and E'. Prove that any isomorphism must be of this form.
[2] Prove that the map in [1] has the following effect upon the coordinates of the curves. If $y^2 = x^3 + ax + b$ is the equation defining E and $y^2 = x^3 + a'x + b'$ is the equation defining E', prove that the effect of the map in Exercise [1] is to take (x, y) to $(c^{-2}x, c^{-3}y)$. (Hint: recall the definition of a and b from Theorem 6.4[1].)

6.2 Elliptic Mahler

Thinking of the way the values $e^{2\pi i\theta}$ parametrize the circle for $0 \leq \theta < 1$, we define, for a polynomial $F \in \mathbb{C}[x]$, the *elliptic Mahler measure* to be

$$m_E(F) = \int_\Pi \log|F(\wp_L(z))|d\mu_\Pi. \tag{6.5}$$

In (6.5), the variable z runs over the canonical fundamental parallelogram Π defined by $L \subset \mathbb{C}$. Explicitly, if L is generated by w_1 and w_2 then

$$\Pi = \{r_1w_1 + r_2w_2 : 0 \leq r_1, r_2 < 1\}.$$

Also, μ_Π denotes the Haar measure on the parallelogram, which coincides with the Lebesgue measure normalized to give area 1 to Π itself. Thus (6.5) is just the usual Riemann integral, normalized to give area 1 to Π. In terms of coordinates, if $z = x + iy \in \Pi$ then $d\mu_\Pi = dxdy/A$ where A is the area of Π. Equivalently, equation (6.5) may be written to reflect the analogy with the classical Mahler measure

$$m_E(F) = \int_0^1 \int_0^1 \log|F(\wp_L(r_1w_1 + r_2w_2))|dr_1dr_2. \tag{6.6}$$

Suppose that E is defined over $\mathbb{Q}$ and the x-coordinate of the point $Q \in E(\mathbb{Q})$ is M/N, a rational number in lowest terms. We might hope for a simple relation between $m_E(Nx - M)$ on the one hand and $\hat{h}(Q)$ on the other. It often happens that

$$m_E(Nx - M) = 2\hat{h}(Q). \tag{6.7}$$

When this does happen, $\log N$ is precisely the total non-archimedean contribution to the canonical height (see Section 5.8). When (6.7) fails it is because of the way the curve (and the point Q) reduce modulo certain primes: the situation is known as 'bad reduction' (see Silverman [Sil86] for the details). It is possible to overcome the failure of (6.7) in several ways. One is to work with an adelic version of the elliptic Mahler measure as in Everest and ní Fhlathúin [EF96], or see Everest [Eve98]. Another is to note that some multiple $Q' = nQ$, $n = n(E)$ a (usually) small constant independent of Q, will satisfy (6.7). If $Q' \neq 0$ then, writing $x(Q') = M'/N'$ we have

$$m_E(N'x - M') = 2\hat{h}(Q').$$

Since $\hat{h}(Q') = n^2\hat{h}(Q)$, this means we can realize the canonical height $\hat{h}(Q)$ as a normalized elliptic Mahler measure

$$2\hat{h}(Q) = \tfrac{1}{n^2}m_E(N'x - M').$$

Example 6.5. For the curve $E : y^2 = x^3 - 36x$ (cf. Example 5.11) we may take $n = 6$ in the discussion above. Thus, if $0 \neq Q \in 6E(\mathbb{Q})$ then (6.7) holds with $x(Q) = \frac{M}{N}$.

More generally, we have the following result.

Theorem 6.6. *Suppose $E : y^2 = x^3 + ax + b$ with $a, b \in \mathbb{Z}$. Let Δ denote the discriminant of E, and let $Q \in E(\mathbb{Q})$ have $x(Q) = M/N$. If, for any prime p dividing Δ, p also divides N, then*

$$m_E(Nx - M) = 2\hat{h}(Q).$$

A proof will be given at the end of this section. In Example 6.5, the only relevant primes are 2 and 3 and it can be checked that for all $0 \neq Q$ in $6E(\mathbb{Q})$, we have $6|N$ where $x(Q) = M/N$.

Example 6.7. Let E denote the curve $y^2 = x^3 + 1$. Suppose that m is an integer. Then m gives rise to an (algebraic) point $Q_m = (m, *)$. The y-coordinate is a quadratic integer. Since $x(2^N Q_m) \in \mathbb{Q}$ for all N, the multiples of Q_m behave like rational points as far as the definition of $\hat{h}(Q_m)$ is concerned. In the discussion above, we may take $n = 2$. This means that for any $m \neq 0, -1$ writing $Q'_m = 2Q_m$ gives

$$m_E(N'x - M') = 2\hat{h}(Q'_m), \quad x(Q'_m) = \frac{M'}{N'} \in \mathbb{Q}.$$

Using the results in Everest [Eve] and Everest and ní Fhlathúin [EF96]

$$2\hat{h}(Q_m) = m_E(x - m) = 2\lambda(Q_m)$$

provided $m \equiv 1$ or $3 \bmod 6$. The condition on $m \bmod 6$ guarantees that Q_m is a point of 'good reduction' for all primes p (in the language of Silverman

[Sil86]). Using the explicit formulæ we see that the only contribution to the canonical height comes from the archimedean component. The calculations in Appendix E use a curve and a point where the only contribution to the global height is archimedean. In that case the elliptic Mahler measure and the canonical height agree (up to a factor of 2).

So we see that the canonical height and the elliptic Mahler measure often coincide. The latter is far more general since it applies to arbitrary complex elliptic curves, enabling a definition of elliptic height to be given in complete generality.

The first and most natural question to ask about a definition like equation (6.5) is whether it exists. In the toral case, we settled this by appealing to Jensen's formula. That is exactly what we propose to do in the elliptic case, by using the elliptic analogue of Jensen's formula. In order to motivate this, suppose α is a zero of F. By Theorem 6.4, there is an $a \in \Pi$ with $\wp_L(a) = \alpha$. We want to show that the following integral exists:

$$\int_\Pi \log |\wp_L(z) - \wp_L(a)| d\mu_\Pi. \tag{6.8}$$

Our claim is that this integral is essentially the logarithm of a classical function, the σ-function.

Let $\sigma_L(z)$ denote the function defined by

$$\sigma_L(z) = z \prod_{0 \neq \ell \in L} \left(1 - \frac{z}{\ell}\right) e^{\frac{z}{\ell} + \frac{1}{2}\left(\frac{z}{\ell}\right)^2}.$$

Exercise 6.6. Prove that

$$\sin z = z \prod_{n=1}^\infty \left(1 - \frac{z^2}{n^2\pi^2}\right) \tag{6.9}$$

for all $z \in \mathbb{C}$.

Thus, if we write

$$\left(1 - \frac{z^2}{n^2\pi^2}\right) = \left(1 - \frac{z}{n\pi}\right)\left(1 + \frac{z}{n\pi}\right),$$

then sin is a product over the points of the lattice $\pi\mathbb{Z}$ in $\mathbb{R}$, which is analogous to the way in which σ_L is a product over the points ℓ in the lattice $L \subset \mathbb{C}$.

Lemma 6.8. *The function $\sigma_L(z)$ is an entire function which vanishes on each of the lattice points $\ell \in L$ and nowhere else.*

Proof. Take logarithms and notice that $\log |\sigma_L(z)| = \Re \log \sigma_L(z)$. Now it is clear that

$$\log\left(1 - \frac{z}{\ell}\right) + \frac{z}{\ell} + \frac{1}{2}\left(\frac{z}{\ell}\right)^2 = -\frac{1}{3}\left(\frac{z}{\ell}\right)^3 - \cdots$$

Thus the right-hand side converges for all z with $|z| < |\ell|$, by comparison with the series $\sum' |\ell|^{-3}$.

One of the many useful properties of this function is that with it, one may factorize elliptic functions (see Lang [Lan78], Silverman [Sil86], [Sil94] and Whittaker and Watson [WW63, Chapter XX]). Such a function is

$$\wp_L(z) - \wp_L(a)$$

and Lemma 6.9 records the precise version of the factorization.

Lemma 6.9. *For any* $z, a \in \mathbb{C}\backslash L$ *with* $z, a \neq 0$,

$$\wp_L(z) - \wp_L(a) = -\frac{\sigma_L(z - a)\sigma_L(z + a)}{\sigma_L^2(z)\sigma_L^2(a)}.$$

Before proving this, let us pause to catch the echo of the parallelogram law from Chapter 5 (see (5.16) and (5.17)). Taking absolute logarithms, for $z, a, z \pm a \neq 0$,

$$\log|\wp_L(z) - \wp_L(a)| + 2\log|\sigma_L(z)| + 2\log|\sigma_L(a)|$$

$$= \log|\sigma_L(z + a)| + \log|\sigma_L(z - a)|. \tag{6.10}$$

This formula can now be used to show that $\log|\wp_L(z) - \wp_L(a)|$ is an integrable function. To see this, note that $\log|\sigma_L(z)/z|$ is a continuous function and therefore integrable on Π. Thus, each of the functions in (6.10) has a logarithmic singularity at worst so they are all integrable. It follows that $\log|\wp_L(z) - \wp_L(a)|$ is integrable because it is a sum of integrable functions. In fact this simple trick will allow us to evaluate the integral in a fairly explicit way.

Proof (of Lemma 6.9). Define

$$h(z) = \frac{\sigma_L^2(z)(\wp_L(z) - \wp_L(a))}{\sigma_L(z + a)\sigma_L(z - a)}.$$

The zeros and poles of h exactly cancel out so this function must be a constant. To determine the constant, let $z \to 0$ and note that $\sigma_L(-a) = -\sigma_L(a)$. Then $h(z) \to -1/\sigma_L^2(a)$, which proves the lemma.

To evaluate the integral in (6.8), we would like to integrate the parallelogram law. There is a problem: the function $|\sigma_L(z)|$ is not periodic with respect to L. In order to rectify this, we need to take a detour. Define

$$\zeta_L(z) = \sum_{0 \neq \ell \in L} \left(\frac{1}{z - \ell} + \frac{1}{\ell} + \frac{z}{\ell^2} \right).$$

This is known as the Weierstrass ζ-function.

Theorem 6.10. [1] $\zeta_L(z)$ *is absolutely convergent for* $z \in \mathbb{C}\backslash L$, *where it defines a meromorphic function with simple poles at all* $\ell \in L$.
[2] *The derivative of* $\zeta_L(z)$ *is* $-\wp_L(z)$.
[3] *For all* $z \in \mathbb{C}\backslash L$ *and all* $\ell \in L$, *we have*

$$\zeta_L(z + \ell) = \zeta_L(z) + \eta_L(\ell),$$

where $\eta_L(\ell)$ *depends upon* $\ell \in L$ *only. The map* $\eta_L : L \to \mathbb{C}$ *is known as the quasi-period map.*
[4] *The quasi-period map is a homomorphism from* L *to* $\mathbb{C}$.
[5] [LEGENDRE RELATION] *Let* $L =< w_1, w_2 >$ *with* $\Im(w_1/w_2) > 0$. *Then*

$$w_1\eta_L(w_2) - w_2\eta_L(w_1) = 2\pi i.$$

Proof. [1] This may be shown in the same way that the convergence of $\wp_L(z)$ was shown.

[2] This follows by differentiating term-by-term.

[3] The function $\zeta_L(z + \ell) - \zeta_L(z)$ must differentiate to zero, using part [2]. Hence it is independent of z.

[4] Consider

$$\eta_L(\ell + \ell') = \zeta_L(z + \ell + \ell') - \zeta_L(z).$$

The right-hand side is

$$\zeta_L(z + \ell + \ell') - \zeta_L(z + \ell) + \zeta_L(z + \ell) - \zeta_L(z) = \eta_L(\ell') + \eta_L(\ell).$$

[5] We use the familiar trick of integrating around Π_β, for a suitably chosen β. Now

$$\int_{\Gamma_\beta} \zeta_L(z)dz = 2\pi i \sum \text{residues of } \zeta_L(z). \tag{6.11}$$

The sum of the residues is 1, since $\zeta_L(z)$ has only a single pole at $z = 0$ with residue 1. On the other hand, the integral in (6.11) can be calculated by combining opposite sides as in the proof of Theorem 6.4[3]. Then using part [3] we obtain the expression

$$w_1\eta_L(w_2) - w_2\eta_L(w_1),$$

completing the proof of part [5].

Now define the normalized version of the σ-function, $\Lambda_L(z)$, by

$$\Lambda_L(z) = e^{-\frac{1}{2}z\eta_L(z)}\sigma_L(z), \quad \lambda_L(z) = -\log|\Lambda_L(z)|,$$

where $\eta_L(z)$ is the quasi-period homomorphism. Note that $\eta_L(z)$ is defined for all z by $\mathbb{R}$-linear extension: if

$$z = r_1w_1 + r_2w_2 \text{ then } \eta_L(z) = r_1\eta_L(w_1) + r_2\eta_L(w_2).$$

As we will see, the failure of $\log|\sigma_L(z)|$ to be periodic has now been rectified.

Theorem 6.11. [1] *The function $\lambda_L(z)$ is well-defined on Π and satisfies the parallelogram law*

$$\log |\wp_L(z) - \wp_L(a)| + \lambda_L(z + a) + \lambda_L(z - a) = 2\lambda_L(z) + 2\lambda_L(a),$$

providing $z, a, z \pm a \neq 0$.
[2] *There is a formula*

$$\lambda_L(a) = -\log |a| + O(1),$$

where the implied constant is uniform, not depending upon a.
[3] [ELLIPTIC JENSEN FORMULA] *For all $0 \neq a \in \Pi$, we have*

$$\int_\Pi \log |\wp_L(z) - \wp_L(a)| d\mu_\Pi = 2\lambda_L(a). \tag{6.12}$$

The reason for naming this the elliptic Jensen formula is part [3]. If a is close to zero in Π then we know that $\log |\wp_L(a)|$ is approximately $-2 \log |a|$ from the Laurent expansion. On the other hand, if $|\wp_L(a)| < 1$ then a is bounded away from zero in Π so $\lambda_L(a) = O(1)$.

Corollary 6.12. *For any $a \neq 0 \in \Pi$,*

$$\int_\Pi \log |\wp_L(z) - \wp_L(a)| d\mu_\Pi = \log \max\{|\wp_L(a)|, 1\} + O(1).$$

It is possible to bound the $O(1)$ term in Corollary 6.12 in quite an explicit fashion: see Lang [Lan78, Chapter 1, Section 8] for a good account. The techniques there use q-expansions which we introduce in the next section. The elliptic Jensen formula makes it likely that some kind of analogue of Schinzel's theorem (Theorem 1.14) holds.

Question 11. Suppose E denotes an arbitrary elliptic curve. Do there exist numbers $\alpha_1, \ldots, \alpha_t \in \mathbb{R}$ with the following property: for any $F \in \mathbb{Z}[x]$ with degree d having real roots,

$$m_E(F) > cd, \qquad c = c(E) > 0,$$

provided $F(\alpha_1) \ldots F(\alpha_t) \neq 0$? Clearly some auxiliary conditions on the zeros of F will be necessary, just as in Schinzel's theorem.

See the papers of Silverman [Sil81] and Hindry and Silverman [HS88] for some related work on lower bounds.

Example 6.13. Let $E : y^2 = x^3 - x$. In Example 6.17[1] we show that $m_E(x) = 0$. We suggest that, with the notation of Question 11, $t = 1$ and $\alpha_1 = 0$. That is, if $F \in \mathbb{Z}[x]$ has degree d and real zeros, then

$$m_E(F) > cd$$

provided $F(0) \neq 0$.

Proof (of Theorem 6.11). To prove the theorem, note that [3] follows by integrating the parallelogram law and using the translation invariance of the measure. This causes three terms to cancel and leaves the formula in [3]. Note how crucial it is that $|\lambda_L(z)|$ is well-defined on Π. Also [2] is clear from the definition of $\Lambda_L(z)$. So it is only [1] which requires proof. To see this, consider

$$\frac{d}{dz} \log \frac{\sigma_L(z+\ell)}{\sigma_L(z)} = \zeta_L(z+\ell) - \zeta_L(z) = \eta_L(\ell).$$

It follows that

$$\sigma_L(z+\ell) = Ce^{\eta_L(\ell)z}\sigma_L(z), \tag{6.13}$$

where C is a constant which does not depend on z. To determine C, first suppose $\ell \notin 2L$. Then $\sigma_L(z)$ does not vanish at $\pm\frac{1}{2}\ell$. Evaluating (6.13) at $z = -\frac{1}{2}\ell$ yields $C = -e^{\frac{1}{2}\eta_L(\ell)\ell}$. If $\ell \in 2L$ then $\sigma_L(z)$ has a simple zero at $\pm\frac{1}{2}\ell$. Letting $z \to -\ell/2$ gives $C = e^{\frac{1}{2}\eta_L(\ell)\ell}$. It follows now that

$$\Lambda_L(z+\ell) = \pm e^{\frac{1}{2}(z\eta_L(\ell) - \ell\eta_L(z))}\Lambda_L(z).$$

Taking the absolute value and using the Legendre relation, we see that $|\Lambda_L(z)|$ is well defined on Π, hence so is its logarithm, as required. The parallelogram law follows from the corresponding result for $\log|\sigma_L(z)|$.

Exercise 6.7. Prove that if $E/\mathbb{C}$ and $0 \neq F \in \mathbb{C}[x]$ has degree d, then

$$m_E(F) \leq h(F) + c_1 d,$$

where $c_1 = \int_\Pi \log^+ |\wp_L(z)| d\mu_\Pi$ (this bound appears in Pinner [Pin98]).

Exercise 6.8. [1] Prove that, for $z \neq 0 \in \Pi$,

$$\wp'_L(z) = -\frac{\sigma_L(2z)}{\sigma_L^4(z)}.$$

If $z \in \Pi$ with $2z \neq 0$ corresponds to (x, y) on E, prove that

$$\lambda_L(2z) = 4\lambda_L(z) + \log|2y|.$$

Notice the similarity to the duplication formula.

[2] Let E denote any complex elliptic curve, associated to the lattice L and with defining equation $y^2 = x^3 + ax + b$, $\Delta = 4a^3 + 27b^2 \neq 0$. Prove

$$\int_\Pi \log|\wp'_L(z)| d\mu_\Pi = \frac{1}{4} \log|16\Delta|.$$

Remark 6.14. If $L = \mathbb{Z}\tau + \mathbb{Z}$, then the local height function λ_L satisfies a form of Poisson's equation

$$\left(\frac{\partial^2}{\partial x^2} + \frac{\partial^2}{\partial y^2}\right)\lambda_L(z) = \frac{2\pi}{\Im(\tau)}$$

for $z = x + \tau y$. See Silverman [Sil94, Exercise 6.5, Chapter VI].

Finally, we can now use the elliptic Jensen formula (Theorem 6.11[3]) to give a proof of Theorem 6.6.

Proof (of Theorem 6.6). Write

$$m_E(Nx - M) = \log N + m_E(x - M/N).$$

Now we find in Silverman [Sil86, Appendix C, Section 18] that $\hat{h}(Q)$ is given as a sum of local heights

$$\hat{h}(Q) = \sum_{p \leq \infty} \lambda_p(Q),$$

where the sum runs over all primes (including infinity). From the explicit formulæ in [Sil86, page 365] we find that

$$2\lambda_\infty(Q) = \lambda(Q)$$

with λ as above. It follows that

$$m_E(Nx - M) = \log N + 2\lambda_\infty(Q)$$

using the elliptic Jensen formula, Theorem 6.11[3]. Finally, for non-archimedean p,

$$\lambda_p(Q) = \tfrac{1}{2} \log^+ |M/N|_p$$

if p does not divide Δ or if $p|N$. Thus

$$\sum_{p \neq \infty} \lambda_p(Q) = \sum_{p \neq \infty} -\tfrac{1}{2} \log |N|_p.$$

Using the product formula (cf. Theorem 5.28), the right-hand side is equal to $\tfrac{1}{2} \log |N|$. We have shown that

$$m_E(Nx - M) = 2 \sum_{p \leq \infty} \lambda_p(Q) = 2\hat{h}(Q),$$

as claimed.

In Silverman [Sil86, page 365], [Sil94, Chapter VI], the explicit formulæ are given for a curve in 'minimal Weierstrass form'. However, it is easily checked that the condition of Theorem 6.6 regarding divisibility of Δ is preserved upon reduction to the minimal case (see Everest [Eve] for the details).

6.3 Elliptic Mahler in Several Variables

For $F \in \mathbb{C}[x_1, \ldots, x_n]$, $m_{E^n}(F)$ begs to be defined by the formula

$$m_{E^n}(F) = \int_{\Pi^n} \log |F(\wp_L(z_1), \ldots, \wp_L(z_n))| d\mu_{\Pi^n}. \tag{6.14}$$

Lemma 6.15. *The generalized elliptic Mahler measure defined by (6.14) exists.*

Proof. Given small $\epsilon > 0$, write

$$m_{E^n}(F) = \int_{|F| < \epsilon} \log |F| + \int_{\epsilon \leq |F| \leq \frac{1}{\epsilon}} \log |F| + \int_{\frac{1}{\epsilon} < |F|} \log |F|. \qquad (6.15)$$

The middle term in (6.15) exists because the integrand is continuous. Using Exercise 6.10[2], together with an integration by parts, the two extreme integrals in (6.15) are bounded in size by a term which is $O(\epsilon^q |\log \epsilon|)$, for some $q > 0$.

It follows from the elliptic Jensen formula (see Corollary 6.12), that, if $F \in \mathbb{C}[x]$ has degree d then

$$m(F) = m_E(F) + O(d). \qquad (6.16)$$

Pinner in [Pin98] shows this kind of relation holds in general. If D denotes any bound on the multi-degree of F, he shows that

$$m(F) = m_{E^n}(F) + O(D).$$

See also Everest and Pinner [EP] for the adelic version of this formula.

We will prove a result about the vanishing of the measure in two variables. It is the analogue of the following result from the toral case: by Jensen's formula,

$$\int_{\mathbb{T}^2} \log |x_1 - x_2| d\mu_{\mathbb{T}^2} = 0.$$

Theorem 6.16. *Let E denote a complex elliptic curve with defining equation $y^2 = x^3 + ax + b$ and discriminant $\Delta = 4a^3 + 27b^2 \neq 0$. Then*

$$\int_{\Pi^2} \log |\Delta^{-\frac{1}{6}}(x_1 - x_2)| d\mu_{\Pi^2} = \int_{\Pi^2} \log |\Delta^{-\frac{1}{6}}(\wp_L(z) - \wp_L(w))| d\mu_{\Pi^2} = 0.$$

Before the proof, we will give a few words of explanation as to why the term $\Delta^{-\frac{1}{6}}$ appears. From Exercise 6.5[1], we know that isomorphisms of E take the form $L \to cL$ for some $0 \neq c \in \mathbb{C}$. Thus, our elliptic Mahler measure is NOT isomorphism invariant. The effect of the isomorphism upon Δ is to map $\Delta \mapsto c^{-12} \Delta$. It follows that the expression

$$\log |\wp_L(z) - \wp_L(a)| - \frac{1}{6} \log |\Delta|,$$

is isomorphism invariant. It is possible to take account of this by defining a normalized height

$$\hat{\lambda}_L(z) = \lambda_L(z) - \frac{1}{12}\log|\Delta|$$

and verifying directly that $\hat{\lambda}_L(z)$ is isomorphism invariant.

For a connection with Arakelov theory, see Lang [Lan88, p. 45], where there is a proof that $\hat{\lambda}_L(z)$ integrates to zero over a fundamental domain. Our proof is based on that one, simplifying a little through the use of the (toral) Jensen formula.

Proof (of Theorem 6.16). From the elliptic Jensen formula (6.12),

$$\int_{\Pi^2} \log|\Delta^{-\frac{1}{6}}(x_1 - x_2)|d\mu_{\Pi^2} = 2\int_{\Pi} \hat{\lambda}_L d\mu_{\Pi}.$$

We go on to show that the right-hand side is zero. In view of the isomorphism invariance, we may replace E by any isomorphic copy. We choose a copy with a lattice L having a basis $L = <1,\tau>$ with $\mathrm{Im}(\tau) > 0$. Let $z = x + iy \in \mathbb{C}$ and write $\tau = t_1 + it_2$ with $t_2 > 0$. Then define $q_\tau = e^{2\pi i\tau}$ and $q_z = e^{2\pi iz}$. Thus our new variable is q_z and we have an isomorphism $\Pi \simeq \mathbb{C}^*/q_\tau^{\mathbb{Z}}$. The new fundamental domain now looks like the annulus below.

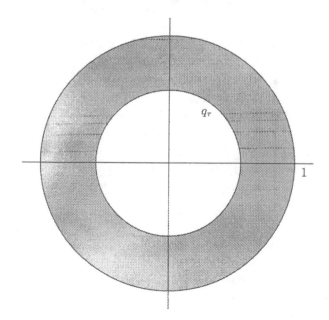

Fig. 6.3. The annulus $\mathbb{C}^*/q_\tau^{\mathbb{Z}}$

The normalized height is given by an explicit formula as follows. The second Bernoulli polynomial $B_2(t) = t^2 - t + \frac{1}{6}$ by definition. Then $\hat{\lambda}_L(z)$ is given by (see Silverman [Sil94, p. 466]),

$$\hat{\lambda}_L(z) = -\log \left| q_\tau^{B_2(y/t_2)/2}(1 - q_z) \prod_{n=1}^{\infty}(1 - q_\tau^n q_z)(1 - q_\tau^n/q_z) \right|. \qquad (6.17)$$

A fundamental domain is given by

$$z = r + s\tau, 0 \le r, s < 1.$$

Write $x = r + st_1, y = st_2$ then

$$\int_{\Pi} \hat{\lambda}_L d\mu_\Pi = \int_0^1 \int_0^1 \hat{\lambda}(r + st_1, st_2)t_2 dr ds.$$

Note firstly that

$$\int_0^1 \int_0^1 \log |q_\tau^{B_2(s)/2}| dr ds = -\pi t_2 \int_0^1 B_2(s) ds = -\pi t_2 \left[\frac{s^3}{3} - \frac{s^2}{2} + \frac{s}{6} \right]_0^1 = 0.$$

Secondly, for $n \ge 1$, consider

$$\int_0^1 \int_0^1 \log |1 - q_\tau^n q_{r+s\tau}| dr ds. \qquad (6.18)$$

Check that $|q_\tau^n q_{s\tau}| = e^{-2\pi t_2(n+s)} < 1$ (which is true since $n \ge 1$). Then the integral in (6.18) vanishes by appeal to Jensen's formula.

Finally, consider

$$\int_0^1 \int_0^1 \log |1 - q_\tau^n q_{-r-s\tau}| dr ds. \qquad (6.19)$$

The same argument as before (although this time $s < 1$ guarantees that $|q_\tau^n q_{-s\tau}| < 1$) shows that the integral in (6.19) vanishes.

In the toral case, we found that the vanishing of the measure for primitive integral polynomials is intimately connected to torsion on the circle (cf. Theorem 1.33). In the elliptic case, we know that the canonical height vanishes precisely on the torsion points of $E(\mathbb{Q})$. In this chapter we are only looking at the archimedean part of the canonical height so we cannot expect such a precise result. See Exercise 6.9[1],[2] and Example 6.17[1],[2] below for examples of torsion points P with $\lambda(P) \neq 0$. In Everest and ní Fhlathúin [EF96], using p-adic analysis to define local Mahler measures, a properly arithmetic measure was worked out. This vanishes precisely when the roots of the polynomial are the x-coordinates of torsion points. We do not know of any points P with $\lambda(P) = 0$ apart from torsion points: is it possible that there are none?

Exercise 6.9. [1] Let E denote the elliptic curve with defining equation $E : y^2 = x^3 + 1$. Let P denote the point $P = (2, 3)$, which has order 6.

Now $2P = (0,1)$, $3P = (-1,0)$ and let the respective values of λ be denoted $\lambda_n = \lambda(nP), n = 1,2,3$. Prove that

$$\lambda_1 = \lambda_2 + \lambda_3, \quad \lambda_2 = \tfrac{1}{3}\log 2, \quad \lambda_3 = \tfrac{1}{4}\log 3.$$

[2] Suppose $P = (x_1, y_1)$ is a point on the elliptic curve E with order 5. Let $x_2 = x(2P)$ and show that

$$\lambda(P) = \tfrac{1}{5}\log\left|\tfrac{x_1 - x_2}{2y_1}\right|.$$

More generally, suppose that P has order N (odd), and write $x_n = x(nP)$ for $n = 1, \ldots, (N-1)/2$. Prove that

$$\lambda(P) = \frac{1}{N}\log\left|\frac{(x_1 - x_2)(x_1 - x_3)\ldots(x_1 - x_{(N-1)/2})}{2y_1}\right|.$$

It is a fact that if E is defined over $\mathbb{Q}$ and $P \in E(\mathbb{Q})$ is non-torsion then the limit

$$2\lambda(P) = \lim_{N\to\infty} \frac{1}{N}\sum_{n=2}^{N}\log|x(nP) - x(P)|$$

exists (see Everest and ní Fhlathúin [EF96]).

Example 6.17. [1] Let E be defined by $y^2 = x^3 - x$. Then

$$m_E(x) = 0, \quad m_E(x \pm 1) = \tfrac{1}{2}\log 2. \tag{6.20}$$

Proof. To see why these formulæ hold, consider the three points of order 2, $(0,0), (\pm 1, 0)$. Label these P, Q and R in any order. Any two of these sum to the third on E. Thus, using the parallelogram law, we obtain three linear equations of the form

$$\log|x(P) - x(Q)| + 2\lambda(R) = 2\lambda(P) + 2\lambda(Q).$$

We can solve these for the three λ-values then Theorem 6.11[3] gives the formulae in (6.20).

[2] Let E be defined by $y^2 = x^3 - \tfrac{1}{3}x + \tfrac{19}{108}$. Then

$$m_E\left(x - \tfrac{2}{3}\right) = 0.$$

Proof. Either of the points $(\tfrac{2}{3}, \pm\tfrac{1}{2})$ is a point of order 5 so this result follows from Exercise 6.9[2].

[3] Given E as in Example [1] above,

$$m_{E^2}(1 + x_1 x_2) = \tfrac{1}{3}\log 2.$$

Proof. In this example $|\Delta| = 4$. If L denotes the associated lattice then, from Theorem 6.16 above, we have

$$\int_{\Pi^2} \log |\wp_L(z) - \wp_L(w)| d\mu_{\Pi^2} = \tfrac{1}{3}\log 2.$$

Let $e \in \Pi$ correspond to the point $(0,0)$. Replacing $\wp_L(w)$ by $\wp_L(w+e)$ does not change the value of the integral, because the w-integral is translation invariant. Using the addition formula, and rearranging the terms, the left-hand side becomes

$$\int_{\Pi^2} \log \left| \frac{1 + \wp_L(z)\wp_L(z)}{\wp_L(z)^2} \right| d\mu_{\Pi^2} = m_{E^2}(1 + x_1 x_2) - 2m_E(x).$$

The last term on the right is zero by Example 6.17[1] above.

[4] Let E be defined by $y^2 = x^3 + 1$. Then

$$m_E(x) = \tfrac{2}{3}\log 2, \quad m_E(x+1) = \tfrac{1}{2}\log 3.$$

These formulæ follow from Exercise 6.9[1].

We conclude with a discussion about a possible elliptic analogue of the 'limit formula', Theorem 3.21, for the toral measure. It should look like this: for any polynomial $0 \neq F \in \mathbb{C}[x_1, x_2]$,

$$\lim_{N \to \infty} \int_{\Pi} \log |F(\wp_L(Nz), \wp_L(z))| d\mu_{\Pi} = \int_{\Pi^2} \log |F(\wp_L(z), \wp_L(w))| d\mu_{\Pi^2}.$$

We would expect to be able to prove this using an analogue of Lawton's method. This requires a result of the following form: given any small $\epsilon > 0$, let

$$B(\epsilon) = \{z \in \Pi : \epsilon < |F(\wp_L(Nz), \wp_L(z))| < 1/\epsilon\}.$$

Then the measure $\mu_{\Pi}(B(\epsilon)) > 1 - C\epsilon^q$, where C and q are independent of N. Exercise 6.10 look at this possibility in special cases.

Exercise 6.10. [1] Given any monic $0 \neq F \in \mathbb{C}[x]$, with degree d prove that

$$\mu_{\Pi}(B(\epsilon)) > 1 - dC\epsilon^q,$$

where C depends on E only and q depends on d.
[2] Given any $0 \neq F \in \mathbb{C}[x_1, \ldots, x_n]$, prove that

$$\mu_{\Pi}(B(\epsilon)) > 1 - D^n C\epsilon^q,$$

where C depends on the coefficients of F and D, q depend on the multi-degree of F.
[3] Let $F(z) = \wp_L(Nz) + \alpha$, for $0 \neq N \in \mathbb{N}$. Prove that $\mu_{\Pi}(B(\epsilon)) > 1 - C\epsilon^{1/2}$, uniformly. The function F is the elliptic analogue of the polynomial $z^N - 1$.

[4] Prove that for any $F \in \mathbb{C}[x]$,

$$m_E(F) = \int_{\Pi} \log |F(\wp_L(Nz))| d\mu_{\Pi}.$$

[5] For $F(z) = \wp_L(Nz) + \wp_L(z) + \alpha$, we conjecture that $\mu_{\Pi}(B(\epsilon)) > 1 - C\epsilon^{\frac{1}{4}}$ uniformly. This is the elliptic analogue of the polynomial $z^N + z + 1$. Assuming this, prove the following limit holds:

$$\lim_{N \to \infty} \int_{\Pi} \log |\wp_L(Nz) + \wp_L(z) + \alpha| d\mu_{\Pi} = \int_{\Pi^2} \log |\wp_L(w) + \wp_L(z) + \alpha| d\mu_{\Pi^2}.$$

6.4 Elliptic Mahler and Periodic Points

In Chapter 1, we looked at Lehmer's original construction of a measure of growth associated to an integer sequence. Let $F \in \mathbb{Z}[x]$ denote a monic integral polynomial. Providing F has no zeros which are roots of unity, we showed (Lemma 1.10) that the following limit holds:

$$\lim_{n \to \infty} \frac{1}{n} \log |\Delta_n(F)| = m(F).$$

In this section, we are going to consider the elliptic analogue of this formula. The reason for doing this will become apparent when we come on to discuss a possible elliptic dynamical system in Section 6.6. We saw in the toral case how $|\Delta_n(F)|$ counts the points of period n in the (toral) dynamical system associated to F. It is natural to conjecture that our elliptic analogue of $|\Delta_n(F)|$ also 'counts' points of period n in some sense.

The starting point for our analogy is to look at the terms $\alpha_i^n - 1$ which comprise $\Delta_n(F)$, each α_i denoting a root of F. Of course this is the cyclotomic polynomial $x^n - 1$ evaluated at α_i. So we now introduce the elliptic analogue of this cyclotomic polynomial.

Suppose E denotes an elliptic curve which is defined by a Weierstrass equation with coefficients in $\mathbb{Q}$. For every $1 \leq n \in \mathbb{N}$, there is a polynomial $\psi_n^2(x)$ of degree $n^2 - 1$ with rational coefficients, whose roots are precisely the $(n^2 - 1)$ x-coordinates of the non-trivial points on E whose order divides n. Moreover the leading coefficient of $\psi_n^2(x)$ is n^2. We refer the reader to Lang [Lan78, Chapter 2] or Cassels' paper [Cas49] for a proof in the classical case, using $\wp$-functions (see also Appendix C). For a more modern treatment, using the generalized Weierstrass equation, see Silverman [Sil94, Exercise 6.4 on page 477]. The notation $\psi_n^2(x)$ is standard for this polynomial. It is the square of a polynomial when n is odd: in general it is the square of a polynomial of the form $y \cdot f$, with square $y^2 f(x)^2 = (x^3 + ax + b)f(x)^2$. See Appendix C for the details. It is true that x-coordinates appear in pairs because $\wp_L(z)$ is even. We prefer to count both a point and its inverse so that we end up integrating over the whole of the parallelogram. It is known that if the defining equation has integral coefficients, then $\psi_n^2(x)$ has integral coefficients also.

Given any monic $F \in \mathbb{Q}[x]$, with roots $\alpha_1, \ldots, \alpha_d$, define

$$E_n(F) = \prod_{i=1}^{d} \psi_n^2(\alpha_i) \text{ for any } 1 \leq n \in \mathbb{N}. \tag{6.21}$$

Theorem 6.18. *Suppose E is an elliptic curve defined over $\mathbb{Q}$. Given any $F \in \mathbb{Q}[x]$ as above, suppose further that no zero of F is the x-coordinate of a torsion point of E. Then the following limit holds:*

$$\lim_{n \to \infty} \frac{1}{n^2} \log |E_n(F)| = m_E(F).$$

Clearly some condition on F, with regard to torsion, is necessary since we also had this in the toral case (there we supposed no zero of F was a root of unity). This formula has an interest in its own right, being the elliptic analogue of our earlier toral result. It takes on more vital interest when we try to find its place in the dynamical scheme of things. For the moment notice that, when E is actually defined over $\mathbb{Z}$, and when $F \in \mathbb{Z}[x]$ is monic, then the numbers $E_n(F)$ are all integers. This is because the polynomial $\psi_n^2(x)$ is integral and to form $E_n(F)$, we take the product of $\psi_n^2(x)$ over algebraic conjugates (the roots of F) which are all algebraic integers. Thus $E_n(F)$ is an algebraic integer and it is fixed under Galois action because it contains all the conjugates of any root of F. Also, note that the numbers $E_n(F)$ form a divisibility sequence, in other words, $E_m(F)|E_n(F)$, whenever $m|n$. This is because if a point P on E has $mP = O$ then $nP = O$ for any $m|n$.

Exercise 6.11. Suppose E is an elliptic curve defined over $\mathbb{Z}$. Then (see Appendix C), ψ_n^2 is an integer polynomial. If $m \in \mathbb{Z}$, prove that $\psi_n^2(m)$ for $n \in \mathbb{N}$ forms a divisibility sequence.

These divisibility sequences satisfy the same recurrence relations as the polynomials ϕ_n and ψ_n^2 (see Appendix C). Morgan Ward [War48] has investigated these sequences in the abstract. He shows that all divisibility sequences satisfying the recurrence relation in Appendix C arise from evaluating either the polynomials ψ_n suitably or the toral counterpart. See also Appendix E for calculations regarding the primality of values of E_n.

Ritt [Rit22], [Rit23] showed a remarkable analogue of Morgan Ward's results characterizing elliptic divisibility sequences in the context of commuting rational maps of the complex plane. Recently, Veselov [Ves91], [Ves92] and others have studied the dynamics of these maps using concepts from physics. The search for elliptic dynamics may well bear fruit within that circle of ideas.

Proof (of Theorem 6.18). Let α denote any algebraic number and suppose $a \in \Pi$ has $\wp_L(a) = \alpha$. We are going to prove the following:

$$\lim_{n \to \infty} \frac{1}{n^2} \log |\psi_n^2(\alpha)| = \int_\Pi \log |\wp_L(z) - \alpha| d\mu_\Pi = 2\lambda_L(a). \qquad (6.22)$$

Notice first of all that the leading coefficient of $\psi_n^2(x)$ contributes nothing to this limit because it produces a term of the shape $O(\log n/n^2)$ which vanishes as $n \to \infty$. If we write ρ for the $n^2 - 1$ non-trivial points on E with order dividing n, then

$$\frac{1}{n^2} \log |\psi_n^2(\alpha)| = \frac{1}{n^2} \sum_\rho \log |x(\rho) - \alpha|. \qquad (6.23)$$

At once it becomes plausible that the right-hand side tends to the integral for large n, because the torsion points densely fill the fundamental parallelogram Π. The only possible hindrance will come from torsion points ρ which are inordinately close to α or to infinity. We will finish the proof by showing this cannot happen. Notice that the latter case exhibits a feature which was absent in the toral case.

Suppose firstly that ρ is close to α. In the toral case, we managed this difficulty by appealing to Baker's theorem on logarithms of algebraic numbers. In the elliptic case, we will appeal to the corresponding statement on 'elliptic logarithms'. Suppose we have a bound of the following shape:

$$|x(\rho) - \alpha| < A/n^B,$$

where A and B are positive constants which depend on E and α only. By Exercise 6.10[1], we see that we have essentially the same bound holding for the pre-images on Π. If $\wp_L(a) = \alpha$ then we have a bound

$$|\rho - a| < A'/n^{B/2}. \qquad (6.24)$$

However, the result of David [Dav95] says that providing the left-hand side is not zero then it is bounded below by C/n^D, where C and D are positive constants which depend on E and α only. We can guarantee that the left-hand side is not zero by our assumption that F has no zeros which are torsion points. Thus, for a suitably large choice of B in (6.24), we cannot have any solutions to the inequality.

Now suppose that ρ is close to infinity. This is easier to deal with because we do not have to appeal to any transcendence result, we can see directly how the proximity of ρ to zero in Π affects the size of $x(\rho)$. Indeed, for ρ close to zero $|x(\rho)|$ is approximately $1/|\rho|^2$. Hence an inequality like

$$|x(\rho) - \alpha| > An^B,$$

will force a similar inequality $|\rho| < A'/n^{B/2}$, where A' and B depend on E and α only.

What this means is that the contribution to the limit in (6.22) from torsion points close to α or infinity is $O(\log n/n^2)$. Obviously this vanishes as $n \to \infty$. Precisely, given any $\epsilon > 0$, we are actually approximating the integral over the region where $\epsilon < |\wp_L(z) - \alpha| < \frac{1}{\epsilon}$. Our explicit bounds mean we are able to choose $\epsilon = A/n^B$ for constants A and B. Then we have shown that the sum is equal to the integral over the region where $\epsilon < |\wp_L(z) - \alpha| < \frac{1}{\epsilon}$ apart from an error which is $O(\log n/n^2) = O(\epsilon^{2/B}|\log \epsilon|)$. This means that (6.23) converges to the integral as claimed.

Exercise 6.12. [1] Prove that

$$n^2 \prod_{n\rho=0} (\wp_L(z) - \wp_L(\rho)) = \left(\frac{\sigma_L(nz)}{\sigma_L(z)^{n^2}}\right)^2. \tag{6.25}$$

[2] Prove that

$$(-1)^{n-1}\{1!2!\ldots(n-1)!\}^2 \frac{\sigma_L(nz)}{\sigma_L(z)^{n^2}} = \det\left(\wp_L^{(i+j-1)}(z)\right)_{1\le i,j\le n-1} \tag{6.26}$$

where $\wp_L^{(k)}(z)$ denotes the kth derivative.

This exercise shows that the hope of realizing the divisibility sequence as a sequence counting periodic points may not be forlorn. Part [2] shows that the terms are essentially determinants, just as their toral counterparts are (cf. Lemma 2.3). The matrix on the right of (6.26) is in 'Hankel form'. There is an extensive classical literature on Hankel matrices: see Gantmacher [Gan59].

6.5 Application to the Division Polynomial

In this section an application of the elliptic Jensen formula to the division polynomial attached to an elliptic curve is presented. Given any complex elliptic curve, define the nth division polynomial by

$$\tau_n(x) = \prod_{\rho:n\rho=0} (x - x(\rho)) \tag{6.27}$$

where ρ runs over all points in Π with $n\rho = 0$. For example, in the notation of Section 6.4, $n^2\tau_n(x) = \psi_n^2(x)$. There is a classical result which states that the coefficients of τ_n are bounded in absolute value by c^{n^2}, where c is a constant depending on the curve E only. This appears in Lang [Lan78, Theorem 3.1 and Theorem 3.2], and is used in transcendence results for elliptic curves (Chapter 9, *ibid.*). David's result in the last section is an example of such an application. The bound on the coefficients for τ_n follows from the elliptic Jensen formula, as we now describe.

Lemma 6.19. *The elliptic Mahler measure of τ_n satisfies*

$$m_E(\tau_n) = O(n^2)$$

uniformly.

Proof. Notice that

$$\frac{1}{n^2} m_E(\tau_n) = \frac{1}{n^2} \sum_{\rho:n\rho=0} \lambda(\rho).$$

Now $\lambda(z) = -\log|z| + g(z)$, where g is a continuous function on Π. It follows that

$$\lim_{n\to\infty} \frac{1}{n^2} \sum_{\rho:n\rho=0} \lambda(\rho) = \int_\Pi \lambda d\mu_\Pi.$$

From Theorem 6.16, the right-hand side is $\frac{1}{6}\log|\Delta|$.

Theorem 6.20. *Writing $h(\sum_{i=0}^d a_i x^i) = \log\max_{0\le i\le d}|a_i|$ for the logarithmic height of a polynomial,*

$$h(\tau_n) = O(n^2).$$

Proof. From the elliptic Jensen formula (6.16),

$$m(F) = m_E(F) + O(d) \tag{6.28}$$

for any polynomial F of degree d. Now by the bound (1.5) (cf. Exercise 1.11),

$$h(F) = m(F) + O(d). \tag{6.29}$$

It follows that

$$\begin{aligned} h(\tau_n) &= m(\tau_n) + O(n^2) \\ &= m_E(\tau_n) + O(n^2) \\ &= O(n^2) \end{aligned}$$

by (6.28), (6.29) and Lemma 6.19 respectively.

Exercise 6.13. Let $E/\mathbb{C}$ denote an elliptic curve and suppose that F is a non-zero polynomial in $\mathbb{C}[x]$ with degree d. Prove that

$$m_E(F') \le m_E(F) + c_1 d$$

for some $c_1 = c_1(E) > 0$.

Question 12. Can you improve the bound to one of the form

$$m_E(F') \le m_E(F) + c_2 \log d$$

for some $c_2 = c_2(E) > 0$?

6.6 Elliptic Mahler and Dynamical Systems

It is remarkable that the subjects of dynamical systems and number theory are connected through the toral Mahler measure. It is natural to wonder whether the same thing happens in the elliptic setting also. Now it seems that a complete theory of 'elliptic dynamical systems' must be more arithmetic in character. In [EW98], this is discussed at some length. With reasonable simplifying assumptions, we can give a good account in the real case.

Given any $\beta \in \mathbb{R}$, we can define an action T on the compact group $\mathbb{T}$ by multiplication by β mod 1. In this way we obtain a dynamical system, though it is not a group endomorphism unless β is an integer. It is possible to compute the topological entropy of this system, written $h(T)$.

Theorem 6.21.
$$h(T) = m(x - \beta) = \log^+ |\beta|$$

This proof would take us too far afield: the topological entropy is really found for an associated shift dynamical system (see Walters [Wal82, Section 7.3]). See Parry [Par60], [Par64] and Hofbauer [Hof78] for background on these maps and their connection to representations of real numbers. For recent work on the dynamics of these and related maps, see the paper of Flatto, Lagarias and Poonen [FL96], [FL97a], [FL97b] and [FLP94].

In Silverman [Sil94, Section 5.2], real elliptic curves are characterized using q-expansions (cf. Section 6.3). In particular, Theorem 2.3 *ibid.* gives an isomorphism

$$E(\mathbb{R}) \cong \mathbb{R}^*/q^{\mathbb{Z}} \tag{6.30}$$

for some $q \in \mathbb{R}$, $0 < |q| < 1$ provided E is defined over $\mathbb{R}$. If the group of complex points $E(\mathbb{C})$ is isomorphic to $\mathbb{C}/L$ with $L = \langle 1, \tau \rangle$ and $\Im(\tau) > 0$ (cf. Theorem 6.4[2] and Exercise 6.1), then q can be taken to be $q_\tau = e^{2\pi i \tau}$. Thus in Exercise 6.1[1] we may take $q = e^{-2\pi}$, which is positive, while in part [2] $q = -e^{-\pi\sqrt{3}}$, which is negative. The right-hand side of (6.30) is isomorphic to $\mathbb{R}/\mathbb{Z}$ if $q < 0$ and to $C_2 \times \mathbb{R}/\mathbb{Z}$ if $q > 0$.

In the first case, $E(\mathbb{R})$ has one connected component and in the second case $E(\mathbb{R})$ has two connected components.

Example 6.22. [1] The curve $y^2 = x^3 + 1$ has $q < 0$ and the real points form one component (see Figure 6.4).
[2] The curve $y^2 = x^3 - n^2 x$, $n \geq 1$ has $q > 0$ and the real points form two components (see Figure 6.5).

Exercise 6.14. Let $E_n(\mathbb{R})$ denote the subgroup of $E(\mathbb{R})$ comprising the points Q with $nQ = 0$. Prove that

$$E_n(\mathbb{R}) = \begin{cases} C_n & \text{if } q < 0 \text{ or } n \text{ is odd,} \\ C_2 \times C_n & \text{if } q > 0 \text{ and } n \text{ is even.} \end{cases}$$

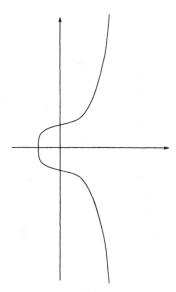

Fig. 6.4. $y^2 = x^3 + 1$

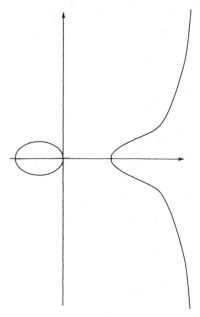

Fig. 6.5. $y^2 = x^3 - n^2 x$

Example 6.23. [1] Let E be given by $y^2 = x^3 - n^2 x$; then $E_3(\mathbb{R}) \cong C_3$ (cf. Exercise 5.8).

[2] Let E be given by $y^2 = x^3 - n^2 x$; then $E_4(\mathbb{R}) \cong C_2 \times C_4$ (cf. Exercise 5.9).

[3] Let E be given by $y^2 = x^3 + 1$; then $E_6(\mathbb{R}) \cong C_6$.

Thus, for E defined over the rationals and $Q \in E(\mathbb{Q})$ there are some indications of how to proceed. Firstly, note that the real analogue of Theorem 6.18 holds: let $\nu_n \in \bar{\mathbb{Q}}[x]$ denote the monic polynomial whose zeros are the x-coordinates of points in $E_n(\mathbb{R})$. Then d_n, the degree of $\nu_n(x)$, is equal to $n - 1$ or $2n - 1$ (cf. Exercise 6.14 above; notice as usual that the identity element does not correspond to a finite x-coordinate).

Theorem 6.24. *If $Q \in E(\mathbb{Q})$ is not a torsion point, then*

$$\lim_{n \to \infty} \frac{1}{d_{2n}} \log |\nu_{2n}(x(Q))| = m_E(x - x(Q)),$$

where m_E denotes the real elliptic Mahler measure

$$m_{E(\mathbb{R})}(x - x(Q)) = \int_{E(\mathbb{R})} \log |x - x(Q)| d\mu_{E(\mathbb{R})}.$$

Notice that if F has real zeros, then $m_{E(\mathbb{R})}(F) = m_{E(\mathbb{C})}(F)$.

Proof. This is the same as the complex case, the point being that $E_n(\mathbb{R})$ is dense in $E(\mathbb{R})$ for large n (this requires n to be even in the case $q > 0$).

Corollary 6.25. *Suppose $E(\mathbb{R})$ has one connected component (so $q < 0$). Given $Q \in E(\mathbb{Q})$ with $m_E(x - x(Q)) > 0$, there is a $\mathbb{Z}$-action T with*
[1] $h(T) = m_E(x - x(Q))$;
[2] $|\mathrm{Per}_n(T)|$ *is asymptotically $|\nu_n(x(Q))|$ provided Q is not a torsion point.*

Proof. We are assuming that $q < 0$ so there is one connected component and $E(\mathbb{R}) \cong \mathbb{T}$. By assumption, $m_E(x - x(Q)) = \log \beta$ for some $\beta > 1$; define $T : \mathbb{T} \to \mathbb{T}$ by $T(x) = \beta x \bmod 1$. Then $h(T) = \log \beta$ and by Flatto, Lagarias and Poonen [FLP94]

$$\lim_{n \to \infty} \frac{1}{n} \log |\mathrm{Per}_n(T)| = \log \beta.$$

In Theorem 6.24 we do not need to assume n is even so $d_n = n - 1$ and

$$\lim_{n \to \infty} \frac{\log |\mathrm{Per}_n(T)|}{\log |\nu_n(Q)|} = 1.$$

Corollary 6.25 should be compared with Corollary 2.8 and Theorem 2.16[1] in Chapter 2.

Question 13. What is the analogue of Corollary 6.25 when q is positive?

An adelic version of Corollary 6.25, using a p-adic version of the β-transformation, appears in [DEM+98].

Question 14. Can you construct a canonical family of maps (parametrized by elliptic curves defined over the rationals and integer polynomials) whose periodic points are related to the formula for $E_n(F)$ in Section 6.4 and whose topological entropies coincide with the elliptic Mahler measures of integer polynomials?

A. Algebra

A.1 Algebraic Integers

Definition A.1. A complex number α is an *algebraic number* if there is a non-zero polynomial $f \in \mathbb{Z}[x]$ for which $f(\alpha) = 0$. If there is a monic polynomial $f \in \mathbb{Z}[x]$ with $f(\alpha) = 0$, then α is an *algebraic integer*.

Example A.2. [1] The notion of algebraic integer generalizes the integers $\mathbb{Z}$. If α is a rational algebraic integer, then $\alpha = \frac{m}{n}$ and so we may take $f(x) = nx - m$ with $n = 1$, showing that α is an integer.
[2] The golden mean $\alpha = \frac{1+\sqrt{5}}{2}$ is an algebraic integer, since $\alpha^2 - \alpha - 1 = 0$.

Theorem A.3. *The set of algebraic integers forms a ring.*

To prove this, we need a useful lemma.

Lemma A.4. *A number $\alpha \in \mathbb{C}$ is an algebraic integer if and only if the additive group generated by all powers $1, \alpha, \alpha^2, \ldots$ is finitely generated.*

Proof. Assume that α is an algebraic integer, so that for some $n > 0$ there is a polynomial

$$\alpha^n + c_{n-1}\alpha^{n-1} + \ldots + c_0 = 0 \tag{A.1}$$

with coefficients $c_j \in \mathbb{Z}$. Let G be the additive group generated by

$$1, \alpha, \ldots, \alpha^{n-1}.$$

By (A.1), $\alpha^n \in G$. Moreover, if $\alpha^m \in G$ for some $m \geq n$, then

$$\alpha^{m+1} = -c_{n-1}\alpha^m - \ldots - c_0\alpha^{m+1-n} \in G,$$

so by induction we see that all powers of α lie in the finitely generated group G.

Conversely, assume that every power of α lies in a finitely generated group Γ. Assume that Γ is generated by n elements. Let G again be the group generated by $1, \alpha, \ldots, \alpha^{n-1}$; since this is a subgroup of Γ it must also be finitely generated; say

$$G = \langle x_i, \ldots, x_n \rangle.$$

Each x_i is a polynomial in α with integral coefficients, so we may write

$$\alpha x_i = \sum_{j=1}^{n} a_{ij} x_j$$

for $j = 1, \ldots, n$. This means that the system of homogeneous equations

$$\begin{bmatrix} a_{11} - \alpha & a_{12} & \cdots & a_{1n} \\ a_{21} & a_{22} - \alpha & \cdots & a_{2n} \\ \vdots & & \ddots & \vdots \\ a_{n1} & a_{n2} & \cdots & a_{nn} - \alpha \end{bmatrix} \begin{bmatrix} x_1 \\ x_2 \\ \vdots \\ x_n \end{bmatrix} = \begin{bmatrix} 0 \\ 0 \\ \vdots \\ 0 \end{bmatrix}$$

has a solution, so the matrix is singular. The equation $\det = 0$ is monic in α and therefore shows that α is an algebraic integer.

Proof (of Theorem A.3). Let α and β be algebraic integers. Let G_α and G_β be the corresponding finitely-generated additive groups generated by all the powers of α and β respectively. Now any power of $\alpha + \beta$ or $\alpha\beta$ is an integer combination of terms of the form $\alpha^i \beta^j$, and therefore lies in the additive group $G_\alpha G_\beta$. On the other hand, if G_α is generated by $x_1, \ldots, x_n$ and G_β is generated by $y_1, \ldots, y_m$, then $G_\alpha G_\beta$ is generated by $\{x_i y_j\}_{i \le n; j \le m}$. It follows that all powers of $\alpha + \beta$ and of $\alpha\beta$ lie in a finitely generated additive subgroup of $\mathbb{C}$, so by Lemma A.4, $\alpha + \beta$ and $\alpha\beta$ are algebraic integers.

Remark A.5. The fact that the algebraic integers form a ring gives the proof that the quantity $\Delta_n(F)$ introduced in (1.1) of Chapter 1 is an integer. If $F \in \mathbb{Z}[x]$ is monic, with roots $\alpha_1, \ldots, \alpha_d$ counted with multiplicities, then recall that

$$\Delta_n(F) = \prod_{i=1}^{d} (\alpha_i^n - 1)$$

for $n \ge 1$. Let Γ be the Galois group of the field extension $\mathbb{Q}(\alpha_1, \ldots, \alpha_d) : \mathbb{Q}$; then the elements of Γ permute the roots of F, so $\Delta_n(F)$ must lie in the fixed field of F, hence

$$\Delta_n(F) \in \mathbb{Q}.$$

On the other hand, the algebraic integers form a ring and each α_i is an algebraic integer, so $\Delta_n(F)$ is an algebraic integer. It follows that $\Delta_n(F)$ must be an integer.

A more surprising result – an easy consequence of Lemma A.4 – is that the roots of monic polynomials with algebraic integer coefficients are also algebraic integers.

Theorem A.6. *Let α be a complex number with the property that*

$$\alpha^n + \beta_{n-1} \alpha^{n-1} + \ldots + \beta_0 = 0$$

for some $n > 0$ and algebraic integers $\beta_0, \ldots, \beta_{n-1}$. Then α is an algebraic integer.

Proof. The coefficients $\beta_0, \ldots, \beta_{n-1}$ generate a subring R of the ring S of algebraic integers. The method of the proof of Lemma A.4 shows that all powers of α lie in a finitely-generated R-submodule M of $\mathbb{C}$, generated by $1, \alpha, \alpha^2, \ldots, \alpha^{n-1}$. By Theorem A.3, each β_i with all its powers lies in a finitely generated additive group G_i with generators $\{x_{ij}\}_{1 \leq j \leq n_i}$ say. It follows that the R-module M lies inside the additive group generated by all the elements

$$x_{1j_1} x_{2j_2} \cdots x_{(n-1)j_{n-1}} \alpha^k$$

for $1 \leq j_i \leq n_i, 0 \leq i \leq n-1, 0 \leq k \leq n-1$, showing that M is finitely generated as an additive group, which implies that α is an algebraic integer by Lemma A.4.

If $K \subset \mathbb{C}$ is a field, then the algebraic integers in K, denoted O_K, are defined to be those algebraic integers that lie in K. It is clear that O_K is a ring.

Example A.7 (1). If $K = \mathbb{Q}(i)$, then $O_K = \mathbb{Z}[i]$, the Gaussian integers.
[2] If $K = \mathbb{Q}(\sqrt{5})$ then the ring of integers is a little larger than might be suspected: by Example A.2[2], $\frac{1+\sqrt{5}}{2}$ is in O_K, and in fact $O_K = \mathbb{Z}[\frac{1+\sqrt{5}}{2}]$.

A.2 Integer Matrices

The natural setting for this result is principal ideal domains: a ring R is a principal ideal domain if every ideal $I \subset R$ is principal: $I = (a)$ for some $a \in R$, where (a) denotes the smallest subring of R containing a. The most important example is the ring $R = \mathbb{Z}$ of integers.

The greatest common divisor of a set $\{a_1, \ldots, a_n\}$ of elements of a principal ideal domain is the uniquely defined (up to multiplication by a unit in R) element d with

$$(d) = (a_1, \ldots, a_n).$$

Theorem A.8. *Let* $a_1, a_2, \ldots, a_n$ *be elements of a principal ideal ring* R, *with greatest common divisor* d_n. *Then there exists an* $n \times n$ *matrix of determinant* d_n *having* $a_1, \ldots, a_n$ *as its first row.*

Corollary A.9. *If* $\mathbf{v} \in \mathbb{Z}^n$ *is a primitive vector, then there is a matrix in* $SL_n(\mathbb{Z})$ *with* $\mathbf{v}$ *as its first column.*

Proof (of Theorem A.8). The result is clear when $n = 2$: since $(d_2) = (a_1, a_2)$, there are elements $c_1, c_2 \in R$ with $d_2 = a_1 c_1 - a_2 c_2$, so we may use the matrix $\begin{bmatrix} a_1 & a_2 \\ c_2 & c_1 \end{bmatrix}$.

Assume the result holds for $n - 1$, and let D_{n-1} be a matrix with $a_1, \ldots, a_{n-1}$ as first row, with determinant equal to d_{n-1}, the greatest common divisor of $a_1, \ldots, a_{n-1}$. Since $(d_n) = (d_{n-1}, a_n)$ we may find $p, q \in R$ such that $p d_{n-1} - q a_n = d_n$. Now define

$$D_n = \begin{bmatrix} & & & & a_n \\ & D_n & & & 0 \\ & & & & \vdots \\ & & & & 0 \\ \frac{a_1 q}{d_{n-1}} & \frac{a_2 q}{d_{n-1}} & \cdots & \frac{a_{n-1} q}{d_{n-1}} & p \end{bmatrix}.$$

It is clear that the determinant of D_n is d_n as required.

Remark A.10. This result is generally attributed to Hermite: in fact he proved more generally that a matrix can always be found with a specified first row $a_1, \ldots, a_n$ and determinant $k_1 a_1 + \ldots + k_n a_n$ for arbitrary $k_1, \ldots, k_n$. Other proofs of Theorem A.8 have been given by Weihrauch (1876) and Hancock (1924). Indeed, Smith (1861) and Stieltjes (1890) proved that if a $p \times n$ matrix of integers is given, $p < n$, and the greatest common divisor of all the $p \times p$ subdeterminants is d, then it is always possible to add $n - p$ rows of integers to the matrix to produce an $n \times n$ matrix with determinant d, and described all the possible forms of the additional rows.

A thorough treatment of integer matrices may be found in Macduffee's book [Mac33].

A.3 Hilbert's Nullstellensatz

In this section we prove the version of the Nullstellensatz needed for the proof of Lemma 5.16 concerning the effect of morphisms on heights. The result is standard but not all texts on algebraic geometry or commutative algebra will have the exact formulation we need. We therefore assume a more standard formulation below and simply derive the form we need using 'Rabinowitsch's trick'.

Suppose k is a field and let $R = k[x_1, \ldots, x_n]$ denote the ring of polynomials over k in n variables. A standard fact from commutative algebra (see for example Reid's notes [Rei88, Chapter 2, Section 3]) is that all ideals in R are finitely generated. Denote by $\bar{k}$ an algebraic closure of k (for the fields we consider, this may be thought of as lying inside the complex numbers $\mathbb{C}$).

Theorem A.11. *Given polynomials $g_1, \ldots, g_t \in R$, suppose that $f \in R$ has the property that*

$$\text{for all } \mathbf{a} \in \bar{k}^n, \ g_1(\mathbf{a}) = \ldots = g_t(\mathbf{a}) = 0 \Rightarrow f(\mathbf{a}) = 0.$$

Then for some $m \geq 1$ and polynomials $h_1, \ldots, h_t \in R$,

$$f^m = g_1 h_1 + \ldots + g_t h_t.$$

Lemma A.12. *Suppose that I is an ideal in R. Then $I = R$ or there is an element $\mathbf{a} = (a_1, \ldots, a_n) \in \bar{k}^n$ such that $f(\mathbf{a}) = 0$ for all $f \in I$.*

Proof. See Reid [Rei88, Theorem 3.10].

Proof (of Theorem A.11). Assume that f is non-zero. Let

$$J = \langle g_1, \ldots, g_t, 1 - yf \rangle \subset k[x_1, \ldots, x_n, y] = R',$$

an ideal in R'. By the Lemma, this must be equal to R' (since there is no element $(a_1, \ldots, a_{n+1}) \in \bar{k}^{n+1}$ which vanishes on all of the generators of J). Thus, for some polynomials $g_0, h'_1, \ldots, h'_t \in R'$

$$1 = g_0(1 - yf) + g_1 h'_1 + \ldots + g_t h'_t.$$

Evaluate at $y = \frac{1}{f}$: the polynomials $h'_1, \ldots, h'_t$ may contain inverse powers of f but only to some bounded degree. Multiplying through by an appropriate power of f now gives

$$f^m = g_1 h_1 + \ldots + g_t h_t$$

for some $h_1, \ldots, h_t \in R$.

B. Analysis

B.1 Stone–Weierstrass Theorem

We have used uniform approximation of continuous functions by trigonometric polynomials several times. The existence of such approximations comes not from Fourier analysis but rather from a very general result, the Stone–Weierstrass theorem. In this section we give a short proof of this general result.

Let X be a compact Hausdorff space, $C(X)$ the space of continuous real-valued functions on X, and write $\|f\| = \sup_{t \in X} |f(t)|$ for the supremum norm on X.

Definition B.1. Let $\mathcal{A}$ be a subset of $C(X)$. Then
$\mathcal{A}$ is an *algebra* if $\mathcal{A}$ is a real vector space and for any $f, g \in \mathcal{A}$, the pointwise product $fg \in \mathcal{A}$,
$\mathcal{A}$ is a *lattice* if, for all $f, g \in \mathcal{A}$ the functions

$$f \vee g = \max\{f, g\}$$
$$f \wedge g = \min\{f, g\}$$

are both in $\mathcal{A}$,
$\mathcal{A}$ *separates points* if for all $x \neq y \in X$ there is a function $f \in \mathcal{A}$ for which $f(x) \neq f(y)$.

Theorem B.2. *Let $\mathcal{A}$ be an algebra which separates points and contains the constant functions. Then every function in $C(X)$ is a limit under the supremum norm of a sequence of elements of $\mathcal{A}$. Equivalently, the closure $\bar{\mathcal{A}}$ of $\mathcal{A}$ under $\| \cdot \|$ is all of $C(X)$.*

Proof. If $f \in \bar{\mathcal{A}}$ then f is continuous so there is a constant K with $\|f\| \leq K$. Then

$$|f| = \sqrt{K^2 - (k^2 - f^2)} = K\sqrt{1 - \left(1 - \frac{f^2}{K^2}\right)}$$

$$= K\left(1 - \sum_{n=1}^{\infty} \frac{1 \cdot 1 \dots (2n-3)}{2 \cdot 4 \cdot 6 \dots (2n)}\left(1 - \frac{f^2}{K^2}\right)^n\right),$$

which is uniformly convergent. It follows that $|f| \in \bar{A}$ since $\bar{A}$ is a uniformly closed algebra. Notice that in proving this we needed to use the fact that the constant function K is in $\bar{A}$.

Now let f and g be functions in $\bar{A}$. Then

$$f \vee g = \tfrac{1}{2}\left(f + g + |f - g|\right),$$
$$f \wedge g = \tfrac{1}{2}\left(f + g - |f - g|\right),$$

so the previous paragraph shows that $\bar{A}$ is a lattice.

Lemma B.3. *If we can find for any $x \neq y \in X$ and $a, b \in \mathbb{R}$ a single function $f_{(x,y)} \in \bar{A}$ such that $f_{(x,y)}(x) = a$ and $f_{(x,y)}(y) = b$, then $\bar{A} = C(X)$.*

Proof. Let $g \in C(X)$ and fix an $\epsilon > 0$. For each pair $x \neq y \in X$ choose a function $f_{(x,y)} \in \bar{A}$ with

$$f_{(x,y)}(x) = g(x), \quad f_{(x,y)}(y) = g(y).$$

Put

$$U_{(x,y)} = \{z \in X : f_{(x,y)}(z) < g(z) + \epsilon\}$$
$$V_{(x,y)} = \{z \in X : f_{(x,y)}(z) > g(z) - \epsilon\}.$$

By construction – and continuity – both are open sets containing x and y.

For fixed y, the family of sets $\{U_{(x,y)}\}_{x \in X}$ is an open cover of X. By compactness, we may find a finite subcover

$$U_{(x_1,y)}, U_{(x_2,y)}, \ldots, U_{(x_n,y)}$$

which covers X. Define

$$h_y = f_{(x_1,y)} \wedge \ldots f_{(x_n,y)} < g + \epsilon$$

by construction. Now

$$h_y > g - \epsilon \text{ on } V_y = V_{(x_1,y)} \cap V_{(x_2,y)} \cap \ldots \cap V_{(x_n,y)},$$

and $\{V_y\}_{y \in X}$ is an open cover of X. By compactness again, we may find a finite subcover

$$V_{y-1}, \ldots, V_{y_m}$$

and now define

$$h = h_{y-1} \vee \ldots \vee h_{y_m}.$$

Then by construction, $g - \epsilon < h < g + \epsilon$ on all of X, so $\|h - g\| < \epsilon$ and g is in the closure of $\bar{A}$.

On the other hand, $\bar{A}$ is closed – so $g \in \bar{A}$.

In order to finish the proof of Theorem B.2 it is enough to check that we can find the function $f_{(x,y)}$ from Lemma B.3. If $x \neq y \in X$ and $a, b \in \mathbb{R}$ then we can certainly find $f \in \bar{A}$ with $f(x) \neq f(y)$. It follows that the matrix $\begin{bmatrix} f(x) & 1 \\ f(y) & 1 \end{bmatrix}$ is non-singular, so we may solve the equations

$$\alpha f(x) + \beta = a,$$
$$\alpha f(y) + \beta = b$$

for α and β. Then we may take $f_{(x,y)} = \alpha f + \beta \in \bar{A}$.

Corollary B.4. *Let $C_{\mathbb{C}}(X)$ denote the complex-valued functions on X. If $A \subset C_{\mathbb{C}}(X)$ is an algebra which separates points, contains the constants and is closed under complex conjugation, then $\bar{A} = C_{\mathbb{C}}(X)$.*

Proof. The real and imaginary parts of the functions in A satisfy the hypotheses of Theorem B.2, so we need only check that the real and imaginary parts of each function can be found. For this it is enough to use complex conjugation:

$$\Re(f) = \tfrac{1}{2}(f + \bar{f}); \quad \Im(f) = \tfrac{1}{2}(f - \bar{f}).$$

Example B.5. [1] Let A be the algebra generated by the functions $\{e^{2\pi i n x}\}_{n \geq 0}$ on the additive circle $\mathbb{R}/\mathbb{Z}$. Then A separates points and contains the constants. However A is not closed under complex conjugation and indeed $e^{-2\pi i x} \notin \bar{A}$. To see this write $f(x) = e^{-2\pi i x}$ and notice that

$$\int_0^1 f \bar{g} \, dx = 0$$

for all $g \in A$. So, by the dominated convergence theorem, $\int_0^1 f \bar{g} \, dx = 0$ for all $g \in \bar{A}$. However, if $e^{-2\pi i x}$ were in $\bar{A}$, this would imply that $\int_0^1 f(x) e^{2\pi i x} \, dx = 0$, which is false.

[2] If $X \subset \mathbb{R}^n$ is any compact set, then the algebra of all restrictions of real polynomials in n variables to X is uniformly dense in $C(X)$.

[3] If A is the algebra generated by the functions $\{e^{2\pi i n x}\}_{n \in \mathbb{Z}}$ then $\bar{A} = C(\mathbb{R}/\mathbb{Z})$.

[4] Similarly, the algebra generated by the functions $\{e^{2\pi i(n_1 x_1 + \ldots + n_k x_k)}\}_{n \in \mathbb{Z}^k}$ is uniformly dense in $C(\mathbb{R}^k/\mathbb{Z}^k)$.

Remark B.6. Weierstrass originally proved that the polynomials are uniformly dense in $C[0,1]$, and the general result was proved later by Stone. A natural question is to ask if the algebra generated by other sets of powers could be dense, and this has a very precise answer. The Müntz–Szasz theorem says that if $0 < \lambda_1 < \lambda_2 < \ldots$, and A is the algebra of functions on $[0,1]$ generated by

$$1, t^{\lambda_1}, t^{\lambda_2}, \ldots$$

then there are only two possibilities. If $\sum_{n=1}^{\infty} \frac{1}{\lambda_n} = \infty$, then $\bar{A} = C[0,1]$, and if $\sum_{n=1}^{\infty} \frac{1}{\lambda_n} < \infty$ then $\bar{A}$ does not contain the function t^{λ} for any $\lambda \notin \{\lambda_n\}_{n \geq 0}$.

This remarkable result shows that whenever a set $\{1, t^{\lambda_1}, t^{\lambda_2}, \ldots\}$ generates a uniformly dense algebra in $C[0,1]$, infinitely many of the functions may be omitted without altering the uniform denseness. Thus $C[0,1]$ does not have minimal sets of functions generating uniformly dense algebras. By contrast, families of orthonormal functions in a Hilbert space (Fourier analysis) generating a dense algebra cannot lose any members without losing the density.

B.2 The Gelfand Transform

The Gelfand transform was used in the proof of Theorem 2.13. We assemble the essential facts about this map here: proofs may be found for example in Rudin [Rud73].

Definition B.7. A (complex) *Banach algebra* is an algebra $\mathcal{B}$ over $\mathbb{C}$ with a norm $\| \cdot \|$, with respect to which it is a Banach space, and such that

$$\|xy\| \leq \|x\|\|y\| \text{ for all } x, y \in \mathcal{B}. \tag{B.1}$$

Example B.8. [1] The complex numbers $\mathbb{C}$ themselves with the usual norm form a Banach algebra.
[2] For any finite set A, the complex-valued functions $A \to \mathbb{C}$ with norm $\|f\| = \max_{a \in A} |f(a)|$ form a Banach algebra isomorphic to $\mathbb{C}^n$.
[3] If $\mathcal{B}$ is any Banach space, then the trivial multiplication $xy = 0$ for all $x, y \in \mathcal{B}$ makes $\mathcal{B}$ into a Banach algebra.
[4] If X is a compact metric space, then the algebra of continuous functions $f : X \to \mathbb{C}$ with norm $\|f\| = \sup_{x \in X} |f(x)|$.
[5] Let S be any Banach space. Then the space $\mathcal{B}$ of bounded linear operators $\phi : S \to S$ with respect to the operator norm

$$\|\phi\| = \sup_{\|s\|_S \leq 1} \|\phi(s)\|_S$$

is a Banach algebra. If $S \cong \mathbb{C}^n$ then $\mathcal{B} = M_n(\mathbb{C})$, the $n \times n$ complex matrices.
[6] A closed subalgebra of a Banach algebra is a Banach algebra.

A *homomorphism* $A \to \mathcal{B}$ of Banach algebras is a linear map that respects the multiplication. Any linear subspace of a Banach algebra that is closed under multiplication is called an *ideal*.

If $\mathcal{B}$ is a Banach algebra in which $xy = yx$ for all x, y then it is a *commutative Banach algebra*.

Theorem B.9. *Let Δ be the set of all complex homomorphisms $\mathcal{B} \to \mathbb{C}$ for a commutative Banach algebra $\mathcal{B}$.*
[1] Every maximal ideal in $\mathcal{B}$ is closed.
[2] Every maximal ideal is the kernel of some $h \in \Delta$, and the kernel of every $h \in \Delta$ is a maximal ideal.

The formula

$$\hat{x}(h) = h(x) \tag{B.2}$$

assigns a function $\hat{x} : \Delta \to \mathbb{C}$ to each $x \in \mathcal{B}$. The *Gelfand topology* on Δ is the weakest topology in which every map $\hat{x}$ is continuous; this makes $x \mapsto \hat{x}$ into a map

$$\hat{} : \mathcal{B} \longrightarrow C(\Delta),$$

called the *Gelfand transform*.

Using Theorem B.9[2] to identify Δ with the space $M(\mathcal{B})$ of maximal ideals in $\mathcal{B}$ (and to induce a topology on $M(\mathcal{B})$ gives another form of the Gelfand transform

$$* : \mathcal{B} \longrightarrow C(M(\mathcal{B})).$$

Theorem B.10. *Let $\mathcal{B}$ be a commutative Banach algebra. Then the space $M(\mathcal{B})$ is a compact Hausdorff space, and the Gelfand transform is a homomorphism from $\mathcal{B}$ onto a subalgebra $\mathcal{B}^*$ of $C(M(\mathcal{B}))$ with*

$$\|x^*\|_\infty \leq \|x\|_\mathcal{B}.$$

C. Division Polynomials

In Chapter 6 we were able to compare the torsion points on the circle with those on an elliptic curve. There is another angle to this: from Galois theory the n nth roots of unity are algebraic integers. The analogous statement for an elliptic curve goes as follows: if E is defined over $\mathbb{Q}$, then the $n^2 - 1$ complex points on E with period dividing n are all algebraic numbers.

The analogy is very close. To get roots of unity, we evaluate the function exp at the transcendental numbers $2\pi i k/n$, $k = 0,\ldots,n-1$ and obtain algebraic numbers. In the elliptic case, we evaluate the function $\wp_L$ at the points $(r_1\omega_1 + r_2\omega_2)/n$ for $0 \le r_1, r_2 < n$ (which are generally transcendental) to obtain algebraic numbers.

It is possible to say more about these algebraic numbers because the polynomial equations satisfied by their x coordinates are readily described; they are elliptic analogues of the polynomials $\frac{x^n-1}{x-1}$. In general they are not algebraic integers although their 'denominators' must divide n^2.

Suppose we have a defining equation $y^2 = x^3 + ax + b$ with $a, b \in \mathbb{Q}$. For a typical point $P = (x, y)$ the addition formula says that

$$x(2P) = \left(\frac{3x^2 + a}{2y}\right)^2 - 2x,$$

so the denominator is $4y^2$. The roots of the equation $y^2 = 0$ are precisely the x coordinates of the points of order 2. This process can be continued: the denominator of $x(3P)$ is

$$\left(3x^4 + 6ax^2 + 12bx - a^2\right)^2,$$

with (in general) four double zeros each of which gives rise to two points on the curve. These $8 = 3^2 - 1$ points are the points whose order is 3, and it is clear that these are all algebraic. The process can be continued by induction.

Define polynomials $\psi_n \in \mathbb{Q}[x, y]$ as follows:

$$\psi_1 = 1$$
$$\psi_2 = 2y$$
$$\psi_3 = 3x^4 + 6ax^2 + 12bx - a^2$$
$$\psi_4 = 4y(x^6 + 5ax^4 + 20bx^3 - 5a^2x^2 - 4abx - 8b^2 - a^3)$$

$$\psi_{2n+1} = \psi_{n+2}\psi_n^3 - \psi_{n-1}\psi_{n+1}^3 \text{ for } n \geq 2$$
$$2y\psi_{2n} = \psi_n(\psi_{n+2}\psi_{n-1}^2 - \psi_{n-2}\psi_{n+1}^2) \text{ for } n \geq 3.$$

Exercise C.1. Prove that $\psi_n \in \mathbb{Q}[x, y]$ for all n.

Now define

$$\phi_n = x\psi_n^2 - \psi_{n+1}\psi_{n-1}.$$

Exercise C.2. By replacing y^2 with $x^3 + ax + b$, prove the following.
[1] $\phi_n(x) = x^{n^2} +$ lower order terms.
[2] $\psi_n(x)^2 = n^2 x^{n^2-1} +$ lower order terms.
[3] $x(nP) = \frac{\phi_n(x)}{\psi_n^2(x)}.$
[4] $\phi_n(x)$ and $\psi_n^2(x)$ are relatively prime polynomials.

This may be interpreted as follows. The $n^2 - 1$ roots of $\psi_n^2(x) = 0$ are the x coordinates of the $n^2 - 1$ points whose order divides n. Evaluating $x(nP)$ at any one of these points gives the point at infinity, which agrees with the choice that $0 \in \Pi$ corresponds to the point at infinity in the curve.

These exercises are standard: see Cassels [Cas49], Lang [Lan78] or Silverman [Sil86, page 105]. With a little more effort one can show that for $a, b \in \mathbb{Z}$, $\psi_n, \phi_n \in \mathbb{Z}[x, y]$.

When E is defined over $\mathbb{Z}$ so that $\psi_n^2 \in \mathbb{Z}[x]$, the greatest common divisor of the coefficients of ψ_n^2 (the *content*) is 2^{2s} if 2^s exactly divides n. For a proof see Lang [Lan78, Theorem 2.2].

It is becoming usual in modern texts on elliptic curves to work with a more general form of defining equation of the form

$$y^2 + a_1xy + a_3y = x^3 + a_2x^2 + a_4x + a_6, a_1, a_3, a_2, a_4, a_6 \in K, \qquad \text{(C.1)}$$

which is the *generalized Weierstrass equation*. There are advantages to working with this equation. For example, it enables a smoother development of the addition formulæ over any base field (without restricting the characteristic). It also enables a simpler development of reduction modulo primes. If the coefficients a_1, a_3, a_2, a_4, a_6 are integers, then the polynomials ϕ and ψ are always primitive.

The disadvantages of the form (C.1) are that the formulæ become more complicated and one loses sight of the classical functions. We particularly wanted to keep the function-theoretic point of view while developing the elliptic Mahler measure. The generalized equation appears in Appendix E, where a curve of this form is used to generate prime values of $E_n(F)$. See also Exercise 5.2 in Chapter 5.

D. Proof of Mahler's Bound for $m(F')$

Let $F \in \mathbb{C}[x]$ have degree d. Mahler proved in [Mah61] that

$$m(F') \leq m(F) + \log d. \tag{D.1}$$

Mahler's proof is difficult, so we present a different proof here, following a cryptic remark by the referee at the end of the paper. The proof uses some ideas from subharmonic functions in complex analysis, and applies a simple result due to Lucas.

Mahler's estimate has been used in several other places, including Szegö's theorem: this states that for any polynomial $H \in \mathbb{C}[x]$,

$$M(H) = \inf_Q \left\{ \left(\int_0^1 |H(e^{2\pi it})Q(e^{2\pi it})|^2 \, dt \right)^{1/2} \right\}, \tag{D.2}$$

where the infimum is taken over all monic polynomials Q. A modern proof is in Lawton [Law75]; in [Deg97] Dégot quantified the result by showing how rapidly the right-hand side of (D.2) taken over all monic Q with degree k converges to $M(H)$ as k increases. Szegö's theorem was generalized by Durand [Dur81], who showed that

$$M(H) = \inf_Q \left\{ \left(\int_0^1 |H(e^{2\pi it})Q(e^{2\pi it})|^p \, dt \right)^{1/p} \right\} \tag{D.3}$$

for any $p > 0$, where the infimum is again taken over all monic polynomials Q.

Returning to the proof of (D.1), let $F(z) = a_d z^d + a_{d-1} z^{d-1} + \ldots + a_0$, with zeros $\alpha_1, \ldots, \alpha_d$, and let $F'(z) = d a_d z^{d-1} + \ldots + a_1$ with zeros $\beta_1, \ldots, \beta_{d-1}$. Then

$$M(F) = |a_d| \prod_{j=1}^d \max\{1, |\alpha_j|\} \tag{D.4}$$

and

$$M(F') = d|a_d| \prod_{j=1}^{d-1} \max\{1, |\beta_j|\}. \tag{D.5}$$

Since $F(z) = a_d \prod_{j=1}^{d}(z - \alpha_j)$,

$$\log F(z) = \log a_d + \sum_{j=1}^{d} \log(z - \alpha_j).$$

Differentiating gives

$$\frac{F'(z)}{F(z)} = \sum_{j=1}^{d} \frac{1}{z - \alpha_j}. \tag{D.6}$$

The result claimed is the following.

Theorem D.1. $M(F') \le dM(F)$.

Proof. First notice that

$$\int_0^1 \log \left| \sum_{j=1}^{d} \frac{1}{e^{2\pi i t} - \alpha_j} \right| dt = \log \frac{M(F')}{M(F)}$$

$$= \log d + \log \prod_{j=1}^{d-1} \max\{1, |\beta_j|\} - \log \prod_{j=1}^{d} \max\{1, |\alpha_j|\}$$

by (D.4) and (D.5). It follows that the theorem is equivalent to either of the statements

$$\log \prod_{j=1}^{d-1} \max\{1, |\beta_j|\} \le \prod_{j=1}^{d} \max\{1, |\alpha_j|\} \tag{D.7}$$

or

$$\int_0^1 \log \left| \sum_{j=1}^{d} \frac{1}{e^{2\pi i t} - \alpha_j} \right| dt \le \log d. \tag{D.8}$$

We shall prove (D.8).

Fix $\alpha_1, \ldots, \alpha_{d-1}$ and define a function $u : \mathbb{C} \to \mathbb{R}$ by

$$u(\alpha_d) = \int_0^1 \log \left| \sum_{j=1}^{d} \frac{1}{e^{2\pi i t} - \alpha_j} \right| dt. \tag{D.9}$$

Notice that u exists everywhere since the only possible singularities in the integrand are of the form log(polynomial). Moreover, u is certainly continuous on $\mathbb{C} \backslash \mathbb{S}^1$.

Now by Rudin [Rud74, Theorem 17.3] the function $\phi_t : \mathbb{C} \backslash \mathbb{S}^1 \to \mathbb{R}$ defined by

$$\phi_t(\alpha_d) = \log \left| \sum_{j=1}^{d} \frac{1}{e^{2\pi i t} - \alpha_j} \right|$$

is subharmonic on $\mathbb{C}\backslash\mathbb{S}^1$ for all $t \in [0, 1)$ (cf. Rudin [Rud74, Definition 17.1]). It follows by Theorem 17.5, *ibid* that

$$\int_0^1 \phi_t(\alpha_d + re^{2\pi is})ds \geq \int_0^1 \phi_t(\alpha_d + r'e^{2\pi is})ds \text{ for all } r' < r$$
$$\to \phi_t(\alpha_d) \text{ as } r' \to 0$$

since ϕ_t is continuous. It follows that ϕ_t is a subharmonic function of α_d for each t. Then

$$\int_0^1 u(\alpha_d + re^{2\pi is})ds = \int_0^1 \int_0^1 \phi_t(\alpha_d + re^{2\pi is})dsdt$$
$$= \int_0^1 \int_0^1 \phi_t(\alpha_d + re^{2\pi is})dtds$$
$$\geq \int_0^1 \phi_t(\alpha_d)dt = u(\alpha_d),$$

so u is subharmonic on $\mathbb{C}\backslash\mathbb{S}^1$.

Now apply the maximum modulus principle for subharmonic functions (cf. [CKP83, p.48]) to deduce that

$$\max_{|\alpha_d|<1} u(\alpha_d) = \max_{|\alpha_d|=1} u(\alpha_d)$$

and

$$\max_{|\alpha_d|>1} u(\alpha_d) = \max_{|\alpha_d|=1} u(\alpha_d).$$

The same argument can be applied to each of the d variables in turn, showing that the expression in (D.8) is maximized when all the zeros have $|\alpha_j| = 1$. Write

$$G(z) = \sum_{j=1}^d \frac{1}{z - \alpha_j},$$

and apply the general version of Jensen's theorem (see [CKP83, Section 2-10]) to G:

$$\int_0^1 \log |G(e^{2\pi it})|dt = \log |G(0)| + \log \left(\prod_{j=1}^d |\alpha_j| / \prod_{j=1}^{d-1} |\beta_j| \right). \qquad (D.10)$$

Now return to the original polynomial. We have

$$F(z) = a_d z^d + \ldots + a_0 = a_d \prod_{j=1}^d (z - \alpha_j),$$

so a_1/a_d is a sum of $d = \binom{d}{1}$ products of $(d-1)$-tuples of the α_j's. It follows that

$$\left|\frac{a_1}{a_d}\right| \le d.$$

On the other hand,

$$F'(z) = da_d z^{d-1} + \ldots + a_1,$$

so

$$\prod_{j=1}^{d-1} |\beta_j| = \left|\frac{a_1}{da_d}\right| \le 1.$$

Thus equation (D.10) reduces to

$$\int_0^1 \log|G(e^{2\pi i t})|dt \le \log|G(0)| \le \log\sum_{j=1}^d \frac{1}{|\alpha_j|} = \log d,$$

which proves (D.8) and hence the theorem.

An alternative proof may be given that avoids the use of the general Jensen theorem. Once the general case has been reduced to all $|\alpha_j| = 1$, the following simple argument (Corollary D.3 below) shows that (D.7) holds directly.

Lemma D.2. [LUCAS, 1874] *Let F be any polynomial. If all the zeros of F lie in a half-plane H, then all the zeros of F' lie in the same half-plane.*

Corollary D.3. *The smallest closed convex set containing all the zeros of a polynomial also contains all the zeros of the derivative of the polynomial.*

Lemma D.2 was proved by F. Lucas in several papers beginning in 1874 [Luc79]. A more accessible treatment with the full history and further results is in Borwein and Erdelyi [BE95, Section 1.3] and Marden [Mar49, Chapter 2].

Proof (of Lemma D.2). Assume without loss of generality that the zeros

$$\alpha_1, \ldots, \alpha_d$$

of F lie to the right of the line $z = a + bt$, $b \ne 0$, $t \in \mathbb{R}$. Then

$$H = \{z \in \mathbb{C} : \Im\left(\frac{z-a}{b}\right) < 0,$$

so for $z \notin H$, $\Im\left(\frac{z-a}{b}\right) \ge 0$. Then

$$\Im\left(\frac{z-\alpha_k}{b}\right) = \Im\left(\frac{(z-a)-(\alpha_k-a)}{b}\right)$$
$$= \Im\left(\frac{z-a}{b}\right) - \Im\left(\frac{\alpha_k-a}{b}\right)$$
$$> 0,$$

so

$$\Im\left(\frac{b}{z-\alpha_k}\right) < 0.$$

Multiplying (D.6) by b and taking imaginary parts gives

$$\Im\left(\frac{bF'(z)}{F(z)}\right) = \sum_{j=1}^{d}\Im\left(\frac{b}{z-\alpha_j}\right) < 0,$$

so for any $z \in \mathbb{C}\backslash H$, $F'(z) \neq 0$.

E. Calculations on $\Delta_n(F)$ and $E_n(F)$

By *Bríd ní Fhlathúin*

The calculations in this appendix were performed using PARI–GP (available from the ftp site `math.ucla.edu` in the directory `/pub/pari/`). The primality testing is probabilistic: the numbers identified below as prime are strong pseudo-primes to ten randomly chosen bases. The notation $\#X$ will be used to denote the number of decimal digits in X.

E.1 Lehmer Primes

We can investigate the present efficacy of Lehmer's methods by using computing tools which were not available to him. We work with the same polynomial for which he exhibited results, $F(x) = x^3 - x - 1$, and test values of the sequence $\Delta_n(F)$ (cf. Equation (1.1)) for primality for high values of n. Some results are given in Table E.1: 41 prime numbers are discovered in the range 1 to Δ_{7727} using Lehmer's method.

n	$\#\Delta_n$
127	16
991	122
2693	330
4967	607
7727	944
23 311	2847

Table E.1. Some n's for which $\Delta_n(x^3 - x - 1)$ is prime.

As noted in Section 1.2, Lehmer also studied Δ_n for reciprocal polynomials (cf. Definition 1.17), the square roots of which may give rise to prime numbers. Using the polynomial

$$G(x) = x^{10} + x^9 - x^7 - x^6 - x^5 - x^4 - x^3 + x + 1$$

prime values are generated, some of which are shown in Table E.2: 108 prime numbers were discovered within the range 1 to $\sqrt{\Delta_{4951}}$ using Lehmer's sequence.

n	$\#\sqrt{\Delta_n}$
379	14
2939	105
4951	175

Table E.2. Some n's for which the square root of $\Delta_n(G)$ is prime.

Following the methods of Lehmer and Pierce, explicit prime factors may be found of all values of Δ_n. They point out [Leh33], [Pie17] that the only possible factors of $\Delta_n(F)$ are the prime powers p^e which are of the form $nx + 1$ for some integer x, where $1 \leq e \leq \deg(F)$. For example,

$$\Delta_{86}(G) = 31\,813\,954\,889 = 173 \times 1033 \times 178\,021.$$

In practice, finding such factors becomes cumbersome for large values of n once one exceeds the limits of whatever inbuilt list of primes (up to $500\,000$ in the case of PARI) may be available.

E.2 Elliptic Primes

One can follow through the elliptic analogue of Lehmer's sequence (see Section 6.4) by conjecturing that the integers $E_n(F)$ (cf. Equation (6.21)) to which division polynomials on an elliptic curve give rise at a specific point may also yield prime numbers. Ward [War48] shows that if one looks at the sequence of values of ψ_n for n prime, every prime p divides no more than one element of the sequence, save in the case where p divides two consecutive terms of ψ_m with $2 \leq m \leq 4$. Calculations indicate that there appear to be further constraints on admissible prime divisors: for example, that if p divides ψ_n, we have $n = \pm 1 \bmod p$. As in the classical case, one would wish to work with a point of small height so as to prevent the rapid growth of the sequence.

Working with a general Weierstrass model of the form (C.1), values of the division polynomials were tested for primality on the curve

$$E : y^2 + y = x^3 - x \tag{E.1}$$

using the point $(0,0)$, which is of height $0.0255...$ (see Silverman [Sil94, Exercise 6.10(a)]). Notice that the curve (E.1) can be transformed by setting $w = y + \frac{1}{2}$ into

$$w^2 = x^3 - x + \tfrac{1}{4}. \tag{E.2}$$

For the point $Q = (0,0)$ on E,

$$\hat{h}(Q) = \tfrac{1}{2}m_E(x) = \tfrac{1}{2}\int_\Pi \log|\wp_L(z)|d\mu_\Pi = 0.0255\ldots$$

where L is the lattice corresponding to the Weierstrass form in (E.2).

n	$\#E_n$
23	6
83	77
101	114
409	1857

Table E.3. Some n's for which E_n is prime on $E : y^2 + y = x^3 - x$.

Results are shown in Table E.3. Concentration of prime values is initially dense (nine values of the sequence in the range 1 to E_{101} are prime), trailing away as the rapid growth of the sequence lessens the constraint on possible prime divisors. The sequence clearly increases considerably more rapidly than its non-elliptic counterpart (it is a quadratic exponential in the height). It is possible that an algebraic point of smaller height would yield improved results.

F. Exercises and Questions

F.1 Hints for the Exercises

1.1[1]. The polynomial $5x^2 - 6x + 5$ has $\frac{3\pm 4i}{5}$ as a root.

1.1[2]. If a degree three integer polynomial has roots $\theta_1, \theta_2, \theta_3$ say, with $|\theta_1| = 1$ and $\theta_2 = \bar{\theta}_1$, then $\theta_1\theta_2\theta_3 = \theta_1\bar{\theta}_1\theta_3 = \theta_3$ so $\theta_3 \in \mathbb{Z}$. This means θ_1 is a zero of a monic integer quadratic, which is impossible.

An example in degree 4 is given by $x^4 + 4x^3 - 2x^2 + 4x + 1$: the zero

$$\sqrt{2} - 1 + i\sqrt{2\sqrt{2} - 2}$$

has unit modulus.

1.2. If n divides m then $m = kn$ for some $k \geq 1$ and

$$\frac{\Delta_{kn}(F)}{\Delta_n(F)} = \prod_{i=1}^{d}\left(1 + \alpha_i^n + \alpha_i^{2n} + \ldots + \alpha_i^{(k-1)n}\right)$$

is a symmetric function of the roots of F, and so is an integer.

1.3[1]. The upper bound is obvious. For the lower bound, argue as in Mahler [Mah60]. Let $\{i_1, \ldots, i_m\}$ denote a (possibly empty) subset of $\{1, \ldots, d\}$. Then

$$M(F) \geq |a_d\alpha_{i_1}\ldots\alpha_{i_m}|.$$

Each coefficient a_{d-m} of F is a sum of $\binom{d}{m}$ terms of the form $\pm a_d\alpha_{i_1}\ldots\alpha_{i_m}$. It follows that

$$|a_0| + \ldots + |a_d| \leq \sum_{m=0}^{d} \sum_{\{i_1,\ldots,i_m\}} |a_d\alpha_{i_1}\ldots\alpha_{i_m}|,$$

where the sum is taken over all possible sets of suffices. Hence

$$L(F) \leq M(F) \sum_{m=0}^{d} \binom{d}{m} = 2^d M(F).$$

1.3[2]. Use $F(x) = x^d$ or $F(x) = (x+1)^d$.

1.3[3]. Arrange the zeros of F so that

$$|\alpha_1| \geq |\alpha_2| \geq \ldots \geq |\alpha_d|.$$

Then

$$
\begin{aligned}
|D(F)|^{1/2} = &\, |\alpha_1 - \alpha_d||\alpha_2 - \alpha_d| \ldots |\alpha_{d-1} - \alpha_d| \\
&\times |\alpha_1 - \alpha_{d-1}||\alpha_2 - \alpha_{d-1}| \ldots |\alpha_{d-2} - \alpha_{d-1}| \\
&\ \vdots \\
&\times |\alpha_1 - \alpha_3||\alpha_2 - \alpha_3| \\
&\times |\alpha_1 - \alpha_2|.
\end{aligned}
$$

It follows by the triangle inequality that

$$|D(F)|^{1/2} \leq |\alpha_1 \ldots \alpha_d|^{d-1} \times 2^{1+2+\ldots+(d-1)} \leq 2^{d(d-1)/2} M(F)^{d-1}.$$

1.4[1]. Let K be a splitting field for F, with $O \subset K$ the ring of integers. For any prime number p, $(p) = pO$ is an ideal in O and $O/(p)$ is a ring of characteristic p. It follows that for any zero α of F, $F(\alpha^p) \equiv F(\alpha)^p \equiv 0$ mod pO. The assumption that F is irreducible and non-cyclotomic implies in particular that it has no cyclotomic factor so the expression cannot be zero.

1.4[2]. Choose a prime $p > 2$ with

$$eL(F) < p < 2eL(F).$$

Such a prime always exists – 'Bertrand's postulate' is the statement that for every integer $n \geq 3$ there is a prime p with $n < p < 2n - 2$. Bertrand verified this for $n < 3 \times 10^6$ and Tchebychef proved it in 1850. A simple proof due to Erdös [Erd32] of the result that there is a prime p with $x < p \leq 2x$ for any real $x > 1$ is in Hardy and Wright [HW79, Theorem 417]. Using [1], we have

$$p^d \leq \left| \prod_{i=1}^d F(\alpha_i^p) \right| \leq M(F)^{pd} L(F)^d,$$

so

$$p \leq M(F)^p L(F).$$

By the assumption on p,

$$p \leq M(F)^p L(F) \text{ so } M(F) \geq \left(\frac{p}{L(F)} \right)^{1/p}.$$

Now think of the function $f : p \mapsto \left(\dfrac{p}{L(F)} \right)^{1/p}$ as p varies through all the reals in $(eL(F), 2eL(F)]$. By logarithmic differentiation,

$$f'(p) = \frac{f(p)}{p^2} \cdot (1 - \log p + \ell(F)) < 0$$

since $\log p > 1 + \ell(F)$ for $p > eL(F)$. Thus the minimum value of $f(p)$ occurs when $p = 2eL(F)$. Using the inequality $e^y > 1 + y$ for $y > 0$, this gives $M(F) \geq (2e)^{1/2eL(F)}$, showing (1.9).

1.5. The first identity comes about by multiplying by $e^{i\pi}$ and noting the rotation-invariance of Lebesgue measure. The identity

$$\int_0^1 \log|e^{2\pi i\theta} - 1|d\theta = \int_0^1 \log|e^{4\pi i\theta} - 1|d\theta$$

may be proved directly: for any integrable function $f : [0,1] \to \mathbb{C}$, $\int_0^1 f(2t)dt = \int_0^1 f(t)dt$. (This may be thought of as follows: the transformation $t \mapsto 2t$ mod 1 preserves the Lebesgue measure on $[0,1]$; see Exercises 1.6[1] and 3.1 for another application of this idea.)

1.6[1]. Note that the zeros of $x^n - \alpha$ are given by $x = \xi\alpha^{1/n}$ for some chosen nth root $\alpha^{1/n}$ of α and ξ running through the n nth unit roots. The result follows from the definition of $m(F)$. Alternatively, change the variable of integration in Lemma 1.8.

1.6[2]. A complete discussion is in Appendix D. To see that the inequality cannot be improved, look at $F(z) = z^m$.

1.7. If $(1 \pm x)^n$ divides F then the degree of F may be made arbitrarily large without increasing $M(F)$. The other two parts are straightforward.

1.8. Let θ denote a zero of F with unit modulus. If $\theta = -1$ then, since F is irreducible, $F(x) = \pm(x + 1)$ which is reciprocal. So we may assume that $\theta \notin \mathbb{R}$, and so $\bar{\theta} \neq \theta$ is also a root. By taking complex conjugates and multiplying through by θ^d, we see that θ also satisfies $F^*(x) = 0$. Since the minimal polynomial of θ is unique up to sign, it follows that $F = \pm F^*$. If $F^*(x) = -F(x)$ then $F(1) = F^*(1) = -F(1)$ so $F(1) = 0$, which is impossible. We deduce that F is reciprocal.

1.9. Start with the polynomial $F\bar{F}$.

1.10[1]. If there are two or more non-reciprocal factors then $M(H) \geq \theta_0^2$, where $\theta_0 = M(x^3 - x - 1)$ by Theorem 1.18 together with the fact that $H(0)H(1) \neq 0$. On the other hand, $x \mapsto x^2 + \frac{1}{x^2}$ is monotone increasing for $x > 1$, so by Theorem 1.22

$$\theta_0^4 + \theta_0^{-4} = 3.398\ldots \leq 3,$$

a contradiction.

1.10[2]. Suppose α is a zero of a reciprocal polynomial and also satisfies $H(\alpha) = 0$; that is

$$\alpha^m + \epsilon_1\alpha^n + \epsilon_2 = 0 \text{ for some } \epsilon_1, \epsilon_2 \in \{-1, +1\}. \tag{F.1}$$

Now α^{-1} is a conjugate of α so

$$\alpha^{-m} + \epsilon_1\alpha^{-n} + \epsilon_2 = 0.$$

Multiplying this by $\epsilon_2\alpha^m$ gives

$$\alpha^m + \epsilon_1\epsilon_2\alpha^{m-n} + \epsilon_2 = 0,$$

which with (F.1) yields

$$\epsilon_1\epsilon_2\alpha^{m-n} = \epsilon_1\alpha^n.$$

If $m \neq 2n$ then α is a root of unity. If $m = 2n$ then $\epsilon_2 = 1$ and α is a root of $x^{2n} \pm x^n + 1$. Now these polynomials divide $x^{6n} - 1$, all of whose zeros are roots of unity.

1.11. Just as in Exercise 1.3[1], a_{d-m} is a sum of $\binom{d}{m}$ terms of the form $a_d\alpha_{i_1}\ldots\alpha_{i_m}$, and each of these is bounded above by $M(F)$. It follows that $|a_{d-m}| \leq \binom{d}{m}M(F)$. A simple induction argument shows that $\binom{d}{m} \leq 2^{d-1}$ so we deduce that

$$H(F) \leq 2^{d-1}M(F).$$

The polynomial $F(x) = x - a$ gives equality.

For the reverse direction, use Lemma 1.8:

$$M(F) = \int_0^1 \log|F(e^{2\pi i\theta})|d\theta \leq \left(\int_0^1 |F(e^{2\pi i\theta})|^2 d\theta\right)^{1/2}$$

by Hardy, Littlewood and Polya [HLP34, Section 6.7]. Now apply Parseval's formula, Lemma 1.20 to see that

$$\left(\int_0^1 |F(e^{2\pi i\theta})|^2 d\theta\right)^{1/2} \leq \sum_{k=0}^d |a_k|^2 \leq (d+1)H(F)^2.$$

It follows that

$$M(F) \leq \sqrt{d+1}\,H(F).$$

Equality can occur, but only if F is constant.

1.12. If x^n or $(1-x)^n$ or $(x^2 - x + 1)^n$ divides F then the degree can be made arbitrarily large while the left-hand side of (1.28) remains unchanged.

1.13. If $F(x) = a\prod(x - \alpha_i)$ then $G(x) = a^2\prod(x - \alpha_i + \alpha_i^2) \in \mathbb{Z}[x]$.

1.14. If p divides $2^q - 1$, then $2^q \equiv 1 \bmod p$, so by Fermat's little theorem q divides $p - 1$. Write $p - 1 = 2kq$ (since $p \neq 2$); then

$$\left(\frac{2}{p}\right) \equiv 2^{(p-1)/2} \equiv 2^{qk} \equiv 1 \bmod p,$$

so $p \equiv \pm 1 \bmod 8$ by standard results on the Legendre symbol (see Ribenboim [Rib95b, Section 2.II]).

1.15. We follow short recent proofs of Bruce [Bru93] (sufficiency) and Rosen [Ros88].

Write $M_p = 2^p - 1$, $\tau = \frac{1+\sqrt{3}}{\sqrt{2}}$, $\bar{\tau} = \frac{1-\sqrt{3}}{\sqrt{2}}$, $\omega = \tau^2 = 2 + \sqrt{3}$, $\bar{\omega} = \bar{\tau}^2 = 2 - \sqrt{3}$. Notice first that $\tau\bar{\tau} = -1$ so $\omega\bar{\omega} = 1$.

Lemma F.1. $S_m = \omega^{2^{m-1}} + \bar{\omega}^{2^{m-1}}$.

Proof. Notice first that $\omega\bar{\omega} = 1$. For $m = 1$ the result is clearly true since $S_4 = 4$; assume it is true for m. Then

$$
\begin{aligned}
S_{m+1} &= S_m^2 - 2 \\
&= \left(\omega^{2^{m-1}} + \bar{\omega}^{2^{m-1}} \right)^2 - 2 \\
&= \omega^{2^m} + \bar{\omega}^{2^m} + (\omega\bar{\omega})^{2^{m-1}} - 2 \\
&= \omega^{2^m} + \bar{\omega}^{2^m},
\end{aligned}
$$

which proves the lemma by induction.

Lemma F.2. *In the ring of algebraic integers, if M_p is prime, then $\tau^{M_p+1} \equiv -1 \bmod M_p$.*

What this means is that $\tau^{M_p+1} + 1 = M_p o$ for some algebraic integer o.

Proof. Write $M_p = q$; then

$$
\left(\sqrt{2}\tau \right)^q = \tau^q 2^{(q-1)/2} \sqrt{2} = \left(1 + \sqrt{3} \right)^q \equiv 1 + 3^{(q-1)/2} \sqrt{3} \bmod q,
$$

since if q is prime the binomial coefficients $\binom{q}{j}$ are divisible by q for $j = 1, \ldots, q-1$. Now $q \equiv -1 \bmod 8$, so $2^{(q-1)/2} \equiv \left(\frac{2}{q} \right) \equiv 1 \bmod q$. Similarly, since $q \equiv 1 \bmod 3$, $3^{(q-1)/2} \equiv \left(\frac{3}{q} \right) \equiv -1 \bmod q$. It follows that

$$
\tau^q \equiv \bar{\tau} \bmod q,
$$

which proves the lemma since $\tau\bar{\tau} = -1$.

Turning to the proof of Theorem 1.29, assume first that M_p is prime. Then by Lemma F.2

$$
\tau^{M_p+1} \equiv -1 \bmod M_p \text{ so } \tau^{2^p} + 1 = \omega^{2^{p-1}} + 1 \equiv 0 \bmod M_p. \tag{F.2}
$$

Multiply the last congruence in (F.2) by $\bar{\omega}^{2^{p-2}}$ and use $\omega\bar{\omega} = 1$ and Lemma F.1 to see that

$$\left(\omega^{2^{p-1}} + 1\right)\bar{\omega}^{2^{p-2}} = (\omega\bar{\omega})^{2^{p-2}}\omega^{2^{p-2}} + \bar{\omega}^{2^{p-2}} = S_{p-1} \equiv 0 \bmod M_p.$$

Conversely, if M_p divides S_{p-1}, then by Lemma F.1

$$\omega^{2^{p-2}} + \bar{\omega}^{2^{p-2}} = RM_p$$

for some integer R. Multiplying by $\omega^{2^{p-2}}$ gives

$$\omega^{2^{p-1}} = RM_p\omega^{2^{p-2}} - 1 \tag{F.3}$$

which on squaring gives

$$\omega^{2^p} = \left(RM_p\omega^{2^{p-2}} - 1\right)^2. \tag{F.4}$$

Assume that M_p is composite, and choose an odd prime q that divides M_p and has $q^2 \leq M_p$. We will need some simple observations from group theory: the *order* of an element x in a finite group G is the least integer $n \geq 1$ with $x^n = 1$.

Lemma F.3. [1] *If X is a set with an associative binary operation for which there is an identity, then the set of invertible elements in X forms a group.* [2] *In a finite group, the order of any element is at most the order of the group, and if $x^r = 1$ then the order of x divides r.*

Proof. These are both easy: for the first, notice that 1 (the identity) is itself an invertible element, so it is enough to check that the set of invertible elements is closed under the binary operation: if x, y are invertible then $(xy)^{-1} = y^{-1}x^{-1}$ so xy is also.

The second is also clear: powers of a single element generate a subgroup which cannot have more elements than the whole group.

Let $X = \{a + b\sqrt{3} : a, b \in \mathbb{Z}/q\mathbb{Z}\}$; then X is a ring with multiplicative identity 1. Let X^* denote the multiplicative group of invertible elements: since 0 is certainly not invertible, there are at most $q^2 - 1$ elements in X^*. Now think of ω as an element of X. Since q divides M_p, $RM_p\omega^{2^{p-2}}$ as an element of X must be 0. So in X, (F.4) means that $\omega^{2^p} = 1$ while (F.3) means that $\omega^{2^{p-1}} = -1$. It follows that ω lies in X^* and has order exactly 2^p (recall that q is odd). It follows from Lemma F.3[2] that $2^p \leq q^2 - 1$. However the assumption was that $q^2 - 1 \leq M_p - 1 = 2^p - 2$, which is a contradiction. It follows that M_p must be prime.

1.16. If $F(x) = \prod(x - \alpha)$ then $F^*(x) = \prod(1 - \alpha x) = F(x)\prod\alpha = F(0)F(x)$. Now $F(0) = \pm 1$ since $F(x)$ divides $x^N - 1$ for some $N \in \mathbb{N}$.

2.1[1]. Let G be the closed subgroup. If G is infinite, then since $\mathbb{T}$ is compact G must have a limit point: for any $\epsilon > 0$ there exist distinct points $g, h \in G$

with $\rho(g, h) < \epsilon$ where ρ is the metric on $\mathbb{T}$ inherited from the usual metric on $\mathbb{R}$ (cf. (2.11) later in Chapter 2). Since the metric is translation invariant, it follows that $\rho(g - h, 0) < \epsilon$, so the multiples of $(g - h)$ lie ϵ-densely in the circle. Since G is closed, it follows that $G = \mathbb{T}$. If G is not $\mathbb{T}$, then we have shown it must be finite. If G has k elements, then $kg = 0$ for all $g \in G$, so each element of G is of the form $\frac{j}{k}$. Since there are exactly k such elements, G must comprise exactly those points.

2.1[2]. Let $\theta : \mathbb{T} \to \mathbb{T}$ be an automorphism. Since $\frac{1}{2}$ is the unique element of order 2 in $\mathbb{T}$, we must have $\theta(\frac{1}{2}) = \frac{1}{2}$. Since $\frac{1}{4}, \frac{3}{4}$ are the only elements of order 4, we must have $\theta(\frac{1}{4}) = \frac{1}{4}$ or $\frac{3}{4}$. Assume that $\theta(\frac{1}{4}) = \frac{1}{4}$. Then, since θ maps intervals to intervals, the interval $[0, \frac{1}{4}]$ is either mapped onto itself or onto all of $[\frac{1}{4}, 1]$. However we know where $\frac{1}{2}$ is sent, so the second possibility cannot occur. It follows that $\theta([0, \frac{1}{4}] = [0, \frac{1}{4}]$. Looking at points with order 8 shows in a similar way that $\theta([0, \frac{1}{8}] = [0, \frac{1}{8}]$ and so on. Thus θ fixes all points of the form $\frac{j}{2^k}$ for all j and k, so by continuity θ is the identity. A similar argument starting with $\theta(\frac{1}{4}) = \frac{3}{4}$ gives $\theta(x) = -x$.

2.1[3]. Let $\theta : \mathbb{T} \to \mathbb{T}$ be a surjective endomorphism. The image of θ is all of $\mathbb{T}$ (indeed by [1] we know that if θ is any non-trivial homomorphism it must be onto). Let G denote the kernel of θ; by continuity G is a closed subgroup so must be finite or all of $\mathbb{T}$. The latter is impossible, so the kernel is the finite group $G = \{\frac{j}{k}\}$ for some $k \geq 1$. Define a new isomorphism $\eta : \mathbb{T}/G \to \mathbb{T}$ by $\eta(t + G) = kt$, and let $\bar{\eta} : \mathbb{T} \to \mathbb{T}/G$ be the isomorphism induced by the property $\theta(\bar{\eta}(t + G)) = \theta(t)$. Then $\bar{\eta}\eta^{-1}$ is an automorphism of $\mathbb{T}$, so is the identity or $x \mapsto -x$. In the first case we deduce that $\theta(x) = \bar{\eta}(x + G) = \bar{\eta}\eta^{-1}(kx) = kx$; in the second case $\theta(x) = -kx$.

2.1[4]. Let $\imath_j : \mathbb{T} \to \mathbb{T}$ for $j = 1, \ldots, d$ be the embeddings of each copy of the circle. If $\theta : \mathbb{T}^d \to \mathbb{T}$ is an endomorphism, then $\theta \circ \imath_j : \mathbb{T} \to \mathbb{T}$ is a homomorphism, so is of the form $x \mapsto k_j x$ for some $k_j \in \mathbb{Z}$. It follows that

$$\theta(x_1, \ldots, x_d) = \theta(\imath_1(x_1) + \imath_2(x_2) + \ldots + \imath_d(x_d))$$
$$= \theta(\imath_1(x_1)) + \ldots + \theta(\imath_d(x_d))$$
$$= k_1 x_1 + \ldots + k_d x_d.$$

It follows that if $\theta : \mathbb{T}^d \to \mathbb{T}^d$ is a homomorphism, then if $\pi_i : \mathbb{T}^d \to \mathbb{T}$ is projection onto the ith coordinate, $\pi_i \circ \theta : \mathbb{T}^d \to \mathbb{T}$ is a homomorphism so by the previous paragraph

$$\pi_i \circ \theta(x_1, \ldots, x_d) = a_{i1} x_1 + \ldots + a_{id} x_d$$

with each $a_{ij} \in \mathbb{Z}$. This shows that every homomorphism is given by an integral matrix.

It remains to establish that if the matrix is non-singular, then the homomorphism is onto. Assume first that $\det(a_{ij}) = 0$. Then the rows of the matrix are linearly dependent over $\mathbb{Q}$ and hence there exist integers $m_1, \ldots, m_d$ not all zero with

$$m_1(a_{11}, \ldots, a_{1d}) + \ldots + m_d(a_{d1}, \ldots, a_{dd}) = (0, \ldots, 0).$$

It follows that each point in the image of the homomorphism T_A has the property that

$$m_1 x_1 + \ldots + m_d x_d = 0$$

so T_A is not onto. If the determinant is not zero, then the matrix maps $\mathbb{R}^d$ onto $\mathbb{R}^d$ and so must define an onto map on the quotient $\mathbb{R}^d/\mathbb{Z}^d$.

2.2. Let $\mathbf{a} = (a_1, \ldots, a_d)$ be a rational vector, then we may choose a single integer N with the property that $a_i = \frac{j_i}{N}$ for integers $j_1, \ldots, j_d$. Since the entries of the matrix A are integers, the set of all points on the torus with coordinates of the form $\frac{j}{N}$ (comprising N^d points) is invariant under T_A, so T_A restricted to this set is a permutation. It follows that $T_A^{N^{d!}} \mathbf{a} = \mathbf{a}$.

Notice that if T_A is only assumed to be a surjective endomorphism, then all we can say is that each point with rational coordinates has a finite orbit: that is, the set $\mathbf{a}, T_A \mathbf{a}, T_A^2 \mathbf{a}, \ldots$ is finite.

Now assume that T_A is ergodic and $T_A^n \mathbf{a} = \mathbf{a}$ for some $n \geq 1$. Then $(T_A^n - I)\mathbf{a} = 0$, and by Lemma 2.2 the endomorphism of the torus $T_A^n - I$ corresponds to a non-singular integer matrix B and is therefore surjective. It follows that every coordinate of $\mathbf{a}$ must be of the form $\frac{j}{\det B}$.

2.3. First notice that $B = PAQ$ for some matrices $P, Q \in SL_d(\mathbb{Z})$ and $A = \operatorname{diag}(e_1, \ldots, e_d)$ with $e_i \in \mathbb{Z}$. Now a half-open parallelepiped $\mathcal{P}$ in $\mathbb{R}^d$ is a set defined by inequalities

$$\mathcal{P} = \{\mathbf{x} \in \mathbb{R}^d : 0 \leq L_i(\mathbf{x}) < 1; i = 1, \ldots, d\}$$

for real, linearly independent linear forms

$$L_i(\mathbf{x}) = \sum a_{ij} x_j,$$

with $\det(a_{ij}) \neq 0$. For example, $\mathcal{F} = [0, 1)^d$ is the half-open parallelepiped corresponding to $L_i(\mathbf{x}) = x_i$. The set $B\mathcal{F}$ is also a half-open parallelepiped. The proof now follows in two steps.

First, if $T \in SL_d(\mathbb{Z})$ then

$$|T\mathcal{P} \cap \mathbb{Z}^d| = |\mathcal{P} \cap \mathbb{Z}^d|.$$

To see this, write each linear form as $\mathbf{x} \mapsto \mathbf{a}_t \mathbf{x}$ where $\mathbf{a}$ and $\mathbf{x}$ are column vectors. Then

$$\mathbf{a}_t \mathbf{x} = \mathbf{a}_t \left(T_t T_t^{-1}\right) \mathbf{x} = (\mathbf{a}_t T_t) \left(T_t^{-1} \mathbf{x}\right).$$

Now $T_t \in SL_d(\mathbb{Z})$ so $T_t \mathbf{x} \in T\mathcal{P} \cap \mathbb{Z}^d$ if and only if $\mathbf{x} \in \mathcal{P} \cap \mathbb{Z}^d$.

Secondly, if $T = \operatorname{diag}(e_1, \ldots, e_d)$ with $e_i \in \mathbb{Z}$ then

$$|T\mathcal{P} \cap \mathbb{Z}^d| \cdot |e_1 \dots e_d| = |\mathcal{P} \cap \mathbb{Z}^d| \cdot |\det(T)|.$$

2.4. If $F = 0$ then (2.14) gives $X_F = \mathbb{T}^\mathbb{Z}$, and the map T_F is the left shift on this. The set of points with period $n \geq 1$ is determined by choosing $(x_0, \dots, x_{n-1}) \in \mathbb{T}^n$ arbitrarily and then using the periodicity to deduce the other coordinates of x. It follows that the group $\mathrm{Per}_n(T_F)$ is isomorphic to $\mathbb{T}^n$. Define a small open set

$$B = \{x \in \mathbb{T}^\mathbb{Z} : \rho(x_k, 0) < 1/k\}.$$

Then $\bigcap_{j=0}^{n-1} T_F^{-1} B$ has measure less than or equal to $(1/n)^n \mu(B)$, so by (2.12) the entropy is infinite.

2.5. It is simplest to prove the following general result (due independently to Rokhlin [Rok64] and Halmos [Hal43]); the proof and the application to the problem at hand require some ideas from Fourier analysis on compact groups (see Rudin [Rud62]).

Lemma F.4. *If X is any compact (metrizable) abelian group, and T : $X \to X$ is any continuous endomorphism of X, viewed as a Haar measure-preserving transformation, then T is ergodic if and only if the trivial character $\chi \equiv 1$ is the only element of $\hat{X}$ with the property that $\chi(T^n x) = \chi(x)$ for all x and some $n \geq 1$.*

Proof. If χ is a non-trivial character with $\chi(T^n x) = \chi(x)$ for all x and some $n \geq 1$, then if n is chosen as small as possible with that property, the function

$$f(x) = \chi(x) + \chi(Tx) + \dots + \chi(T^{n-1}x)$$

is a continuous function on X which is invariant under T but is not almost everywhere equal to a constant (since it is a finite sum of orthogonal characters). It follows that T cannot be ergodic.

Conversely, assume that $\chi(T^n x) = \chi(x)$ for all x and some $n \geq 1$ implies that χ is trivial, and let $f \in L^2(X)$ be invariant under T: $f(Tx) = f(x)$ a.e. Expand f as a Fourier series,

$$f(x) = \sum_{\chi \in \hat{X}} c_\chi \chi(x),$$

with $\sum_{\chi \in \hat{X}} |c_\chi|^2 < \infty$ by Parseval's formula (this is proved in Katznelson [Kat76, Lemma 5.4]). Since f is T-invariant,

$$\sum_{\chi \in \hat{X}} c_\chi \chi(Tx) = \sum_{\chi \in \hat{X}} c_\chi \chi(x),$$

so that if $\chi, \chi \circ T, \chi \circ T^2, \dots$ are all distinct, their coefficients must be equal and hence must be zero. So if some $c_\chi \neq 0$, we must have $\chi(T^p x) = \chi(x)$

for all x and some $p \geq 1$. By assumption this forces χ to be trivial, so the only non-zero coefficient in the Fourier expansion of f is that of the constant term, showing that f is constant and T is ergodic.

Applying this to the case of T_F acting on X_F, first note that the group of characters of X_F is isomorphic to $\mathbb{Z}[x^{\pm 1}]/F \cdot \mathbb{Z}[x^{\pm 1}]$ (since the character group of $\mathbb{T}^{\mathbb{Z}}$ is isomorphic to $\mathbb{Z}[x^{\pm 1}]$ and the subgroup X_F is simply those elements annihilated by the polynomials $x^k F(x)$ for all $k \in \mathbb{Z}$). Under this isomorphism, the dual of the map T is multiplication by x.

So if T_F is non-ergodic, by Lemma F.4 there must be a polynomial $H \in \mathbb{Z}[x^{\pm 1}]$ that is not divisible by F with the property that $x^n H(x) - H(x) \in F \cdot \mathbb{Z}[x^{\pm 1}]$ for some $n \geq 1$. Choose a zero α of F with $H(\alpha) \neq 0$: then $\alpha^n H(\alpha) = H(\alpha)$ so α must be a root of unity.

Conversely, assume that T_F is ergodic. Then by Lemma F.4, for any polynomial H, $x^n H(x) - H(x) \in F \cdot \mathbb{Z}[x^{\pm 1}]$ for some $n \geq 1$ implies that $H \in F \cdot \mathbb{Z}[x^{\pm 1}]$. It follows that multiplication by $(x^n - 1)$ is injective on $\mathbb{Z}[x^{\pm 1}]/F \cdot \mathbb{Z}[x^{\pm 1}]$ for every $n \geq 1$; the same is true after evaluating at $x = \alpha$ for any zero of F, so F cannot have any unit root zeros.

2.6. If $F(\alpha) = 0$ and $|\alpha| = 1$, then defining $x_k = \Re(\alpha^k)$ defines a point in X_F for any $\epsilon > 0$: if $y = (y_k)$ is defined by $y_k = \alpha^k$ then $y \in X_F$ since

$$a_0 \alpha^k + a_1 \alpha^{k+1} + \ldots + a_d \alpha^{k+d} = \alpha^k F(\alpha) = 0 \qquad (\text{F.5})$$

for all k, and in equation (F.5) we may take the real part and multiply by any $\epsilon > 0$. The existence of a point in X_F which is small in every coordinate shows that the action cannot be expansive.

2.7[1]. Recall from calculus that

$$\sum_{n=1}^{\infty} \frac{x^n}{n} = -\log(1 - x)$$

for $|x| < 1$. It follows that

$$\exp \sum_{n=1}^{\infty} \frac{x^n}{n} = \frac{1}{1 - x}.$$

Applying this to the case $F(x) = a$, Lemma 2.14 shows that

$$|\mathrm{Per}_n(T_F)| = |a|^n,$$

so

$$\zeta_{T_F}(z) = \frac{1}{1 - |a|z}.$$

Strictly speaking, $\frac{1}{1-|a|z}$ is the meromorphic extension of the zeta function, which coincides with it where the zeta function converges.

2.7[2]. Let $F(x) = x^2 - 3x + 1$; the zeros of F are $\lambda = \frac{3}{2} + \frac{\sqrt{5}}{2} > 1$ and $\mu = \frac{3}{2} - \frac{\sqrt{5}}{2} \in (0, 1)$. It follows that

$$\mathrm{Per}_n(T_F) = (\lambda^n - 1)(1 - \mu^n) = \lambda^n - (\lambda\mu)^n + \mu^n - 1 = \lambda^n + \mu^n - 2,$$

so

$$\zeta_{T_F}(z) = \frac{(1 - z)^2}{(1 - \lambda z)(1 - \mu z)}.$$

2.7[3]. The radius of convergence of the zeta function is R, where

$$\frac{1}{R} = \limsup_{n \to \infty} \left(|\mathrm{Per}_n(T_F)|/n\right)^{1/n},$$

and by Theorem 2.16[1], it follows that

$$R = \exp(-h(T_F)).$$

Thus the smallest real pole of the zeta function of T_F is at $\exp(-h(T_F))$.

3.1. Using the Euclidean algorithm, we may write A as a product of integer matrices which are of the the following three types: (i) upper triangular with 1's on the diagonal, (ii) lower triangular with 1's on the diagonal, and (iii) diagonal. Now m is clearly invariant under transformations of the first two types by translation invariance of the integral. It is also invariant under the third by the method of Exercise 1.6[1].

An alternative proof is to note that $\mathbf{x} \mapsto A\mathbf{x}$ is a Lebesgue measure-preserving transformation of the torus (cf. Walters [Wal82, Chapter 1]).

A 'dynamical' proof, showing that the entropies of the corresponding dynamical systems coincide, is in [War96].

3.2. Write $F(\mathbf{x}) = F_{\mathbf{z}}(x_i)$ where $\mathbf{z} = (x_1, \ldots, x_{i-1}, x_{i+1}, \ldots, x_n)$. Then apply the inequality in one variable

$$m(\partial F/\partial x_i) = \int_{\mathbf{z}} \left(\int \log |\partial F/\partial x_i| dx_i \right) d\mathbf{z}$$
$$\leq \int_{\mathbf{z}} \left(m(F_{\mathbf{z}}(x_i)) + \log d_i \right) d\mathbf{z}$$
$$= m(F) + \log d_i.$$

3.3. These may be proved using exactly the same methods as Exercises 1.3 and 1.11. The details are in Mahler's paper [Mah62].

3.4[1]. Solve $x_0 + \prod_{j=1}^{n}(1 + x_j) = 0$ for x_0 and apply Jensen's formula.

3.4[2]. By the double-angle formula,

$$|1 + e^{2\pi i\theta}|^2 = |1 + \cos 2\pi\theta + i \sin 2\pi\theta|^2$$

$$= 1 + 2\cos 2\pi\theta + \cos^2 2\pi\theta + \sin^2 2\pi\theta$$
$$= 2(1 + \cos 2\pi\theta) = 4\cos^2 \pi\theta.$$

3.4[3]. The second integral follows easily from the quoted standard integral (which is in turn a consequence of basic results about the dilogarithm function).

For the first integral, notice that the vanishing of this integral is the essential step in Jensen's formula:

$$\int_0^1 \log|2\cos\pi\theta|d\theta = \int_0^1 \log|e^{\pi i\theta} + e^{-\pi i\theta}|d\theta$$
$$= \int_0^1 \log|e^{2\pi i\theta} + 1|d\theta$$
$$= m(x + 1) = 0.$$

We are grateful to Mark Cooker for pointing out the following alternative proof of the first integral (and therefore of Jensen's formula also). First notice that substituting $t = \frac{\pi}{2} - v$ shows that

$$\int_0^{\pi/2} \log\cos v\,dv = \int_0^{\pi/2} \log\sin t\,dt.$$

Now the Fourier expansion of $\log|2\sin\left(\frac{x}{2}\right)|$ (see Cooker [Coo98, Equation (10)] for example) is

$$\sum_{n=1}^{\infty} \frac{1}{n}\cos nx = -\log\left|2\sin\left(\frac{x}{2}\right)\right| \qquad (\text{F.6})$$

on $(0, \pi]$ for $\sin\left(\frac{x}{2}\right) \neq 0$. Simple convergence arguments justify the following steps. First integrate (F.6) from 0 to x, and then substitute $x = 2t$ to obtain

$$\sum_{n=1}^{\infty} \frac{1}{n^2}\sin nx = -\int_0^x \log\left|2\sin\left(\frac{x}{2}\right)\right|dx$$
$$= -x\log 2 - 2\int_0^{x/2} \log|\sin t|dt.$$

When $x = \pi$ the left-hand side is zero, so

$$0 = -\pi\log 2 - 2\int_0^{\pi/2} \log|\sin t|dt,$$

and since $\sin t \geq 0$ in the range of the integral, it follows that

$$\int_0^{\pi/2} \log\sin t\,dt = -\frac{\pi}{2}\log 2.$$

which gives the result by symmetry about $\frac{\pi}{2}$.

3.5. Write $F(\mathbf{x}) = a\left(1 + \frac{G(\mathbf{x})}{a}\right)$. Then $m(F) = \log|a| + m\left(1 + \frac{G(\mathbf{x})}{a}\right)$. Since $\left|\frac{G(\mathbf{x})}{a}\right| < 1$, $\log\left(\frac{G(\mathbf{x})}{a}\right)$ may be expanded as a Taylor series uniformly convergent in $\mathbf{x}$. Now $\log\left|1 + \frac{G(\mathbf{x})}{a}\right| = \Re\log\left(1 + \frac{G(\mathbf{x})}{a}\right)$; interchanging the sum and the integral gives the result since the integrals of the individual summands all vanish.

3.6. By Jensen's formula, it is enough to evaluate

$$I = \int_0^1 \int_0^1 \log\max\{|1 + e^{2\pi i\theta_1}|, |1 + e^{2\pi i\theta_2}|\}d\theta_1 d\theta_2.$$

After a change of variables

$$
\begin{aligned}
I &= \frac{1}{4\pi^2}\int_0^{2\pi}\int_0^{2\pi} \log\max\{|1 + e^{is}|, |1 + e^{it}|\}dsdt \\
&= \frac{1}{\pi^2}\int_0^\pi\int_0^\pi \log\max\{|1 + e^{is}|, |1 + e^{it}|\}dsdt \text{ by symmetry} \\
&= \frac{2}{\pi^2}\int_0^\pi \log|1 + e^{it}|\int_t^\pi dsdt \\
&= \frac{2}{\pi^2}\int_0^\pi (\pi - t)\log|1 + e^{it}|dt \\
&= -\frac{2}{\pi^2}\int_0^\pi t\log|1 + e^{it}|dt
\end{aligned}
$$

since $\int_0^\pi \log|1 + e^{it}|dt = 0$. In the range $0 \le t < \pi$ expand the logarithm as usual,

$$
\begin{aligned}
I &= \Re\frac{2}{\pi^2}\int_0^\pi t\sum_{n=1}^\infty \frac{te^{itn}}{n}dt \\
&= \Re\frac{2}{\pi^2}\sum_{n=1}^\infty \frac{1}{n}\int_0^\pi te^{itn}dt \\
&= \frac{2}{\pi^2}\sum_{n=1}^\infty \frac{1}{n^3}\left(e^{in\pi} + 1\right) \\
&= \frac{4}{\pi^2}\sum_{n=1}^\infty \frac{1}{(2n-1)^3} \\
&= \frac{4}{\pi^2}\left(\zeta(3) - \sum_{n=1}^\infty \left(\frac{1}{2n}\right)^3\right) \\
&= \frac{4}{\pi^2}\zeta(3)\left(1 - \frac{1}{8}\right) \\
&= \frac{7}{2\pi^2}\zeta(3).
\end{aligned}
$$

3.7[1]. See Smyth [Smy81a, Corollary 2].

3.7[2]. Multiplying by a monomial simply translates the Newton polygon. A polynomial is of the stated form if and only if its Newton polygon lies on a line orthogonal to $(-d, c)$.

3.8. Use the change of variable method in the proof of Theorem 3.19 to transform F into the shape of Exercise 3.5.

3.9. By the Stone–Weierstrass theorem, it is enough to find continuous functions P and Q with the stated properties. Since ϕ is Riemann-integrable it is bounded above by R say, and there is a finite collection of rectangles $\{A_1, \ldots, A_n\}$ with the property that

$$\sum_{i=1}^{n} \operatorname{area}(A_i) \cdot \sup_{(x_1, x_2) \in A_i} \{\phi(x_1, x_2)\} - \int \phi d\mu_{K^2} < \delta/4.$$

The word 'rectangle' means a set of the form

$$\{(e^{2\pi i t_1}, e^{2\pi i t_2}) : t_i \in [a_i, b_i] \subset \mathbb{T}\}.$$

Now define

$$\bar{Q}(x_1, x_2) = \begin{cases} \sup_{(x_1, x_2) \in A_i} \{\phi(x_1, x_2)\} & \text{if } (x_1, x_2) \in A_i \backslash \bigcup_{j \neq i} A_j, \\ R & \text{if not.} \end{cases}$$

Then $Q \geq \phi$ and their integrals are within $\delta/4$. Now approximate $\bar{Q}$ from above by a continuous function Q which is equal to $\bar{Q}$ except very close to the boundary of each rectangle A_i, and fills in continuously to reach the value R on the boundary. This can be done while keeping $\int Q d\mu_{K^2} - \int \bar{Q} d\mu_{K^2} < \delta/4$, which gives a continuous Q with $Q \geq \phi$ and $\int Q d\mu_{K^2} - \int \phi d\mu_{K^2} < \delta/2$.

Repeating the argument from below (or simply repeating the argument for $-\phi$) gives P.

3.10. Notice that $\log|F| = \phi_\epsilon \log|F| + (1 - \phi_\epsilon) \log|F|$. Then by the triangle inequality

$$\limsup_{N \to \infty} \left| \int \log|F(x, x^N)| d\mu_K - \int \log|F| d\mu_{K^2} \right|$$

$$\leq \limsup_{N \to \infty} \left| \int \phi_\epsilon(x, x^N) \log|F(x, x^N)| d\mu_K - \int \phi_\epsilon \log|F| d\mu_{K^2} \right|$$

$$+ \left| \int (1 - \phi_\epsilon(x, x^N)) \log|F(x, x^N)| d\mu_K \right| + \limsup_{N \to \infty} \left| \int (1 - \phi_\epsilon) \log|F d\mu_{K^2} \right|.$$

The first term is zero since the integrand is continuous and so Lemma 3.22 applies. The third term is non-zero but converges to zero as $\epsilon \to 0$ since $\log|F|$ is Riemann-integrable. The second term converges to zero as $\epsilon \to 0$ by the argument used at the end of the proof of Theorem 3.21.

3.11. The upper bound is trivial. For the lower bound, show that there is a region of measure $\gg \frac{1}{d}$ around each of the dth roots of 1 where the inequality holds.

3.12[1]. Since $F(0)F(-1)F(1) \neq 0$, if F is cyclotomic it must be reciprocal and F is clearly not reciprocal unless $d = 2$.

3.12[2]. This is similar: if the zeros of an integral polynomial F are all in $\{0\} \cup K$ then F is a monomial times a reciprocal or anti-reciprocal polynomial (that is, $F(z) = z^\ell H(z)$ with $H^* = \pm H$). It is clear that F_d cannot satisfy this condition for large d.

3.13. By (1.14)), if $m(F(z, z^d)) \neq 0$ then $m(F(z, z^d)) > C(L(F(z, z^d)))$. Since $L(F(z, z^d))$ is uniformly bounded above (for large d $L(F(z, z^d))$ is constant), it follows that $m(F(z, z^d)) > C$ uniformly in d for some C. On the other hand, if $m(F(z, y)) = 0$, then by Theorem 3.21 we must have $m(F(z, z^d)) < C$ for d large enough. It follows that we must have $m(F(z, z^d)) = 0$ for all large d.

3.14. This follows at once from Exercise 1.16.

4.1. Write $F(\mathbf{x}) = \sum c_\mathbf{n} \mathbf{x^n}$, and $L(F) = \sum |c_\mathbf{n}|$ for the sum of the absolute value of the coefficients of F (cf. Remark 1.5 in the one-dimensional case). Define

$$\epsilon = \frac{1}{10L(F)}, \tag{F.7}$$

and assume that no n-tuple $(\alpha_1, \ldots, \alpha_n) \in K^n$ has $F(\alpha_1, \ldots, \alpha_n) = 0$. Claim first that

$$B = \{x \in X_F : \rho(x_{(0,\ldots,0)}, 0) < \epsilon\}$$

is an expansive neighbourhood for T_F. If not, then there is a point $x = (x_\mathbf{n}) \in X_F \backslash \{0\}$ with the property that

$$x \in \bigcap_{\mathbf{n} \in \mathbb{Z}^n} T_F^\mathbf{n}(B) = \{z \in X_F : \rho(z_\mathbf{k}, 0) < \epsilon \text{ for all } \mathbf{k} \in \mathbb{Z}^n\}. \tag{F.8}$$

Let ℓ_∞ denote the Banach space of bounded real-valued functions $(y_\mathbf{n})_{\mathbf{n} \in \mathbb{Z}^n}$. Since $\epsilon < \frac{1}{10}$, there is a unique point $y \in \ell_\infty$ with $y_\mathbf{n} = x_\mathbf{n} \bmod 1$ for all $\mathbf{n} \in \mathbb{Z}^n$. The choice (F.7) of ϵ implies that

$$\left| \sum c_\mathbf{n} y_{\mathbf{k}+\mathbf{n}} \right| < \epsilon L(F)$$

$$< \frac{1}{10}$$

so

$$\sum c_\mathbf{n} y_{\mathbf{k}+\mathbf{n}} = 0 \tag{F.9}$$

for all $\mathbf{k} \in \mathbb{Z}^n$.

The shift $\mathbb{Z}^n$-action σ defined by

$$(\sigma_{\mathbf{k}}(z))_{\mathbf{n}} = z_{\mathbf{k}+\mathbf{n}} \text{ for all } k$$

is an isometric action (that is, each $\sigma_{\mathbf{k}}$ is an isometry of ℓ_∞). Let

$$S = \{z \in \ell_\infty : \sum c_{\mathbf{n}} z_{\mathbf{n}+\mathbf{k}} = 0 \text{ for all } \mathbf{k} \in \mathbb{Z}^n\};$$

by (F.9) the point y lies in S, so S is a non-trivial closed linear subspace of ℓ_∞. The space $\mathcal{B}$ of bounded linear operators $S \to S$ is a Banach algebra; let $\mathcal{A} \subset \mathcal{B}$ be the Banach subalgebra generated by the set $\{\sigma_{\mathbf{n}}\}_{\mathbf{n} \in \mathbb{Z}^n}$. The Gelfand transform $* : \mathcal{A} \to \mathcal{C}(M(\mathcal{A}))$ from $\mathcal{A}$ to the Banach algebra of continuous $\mathbb{C}$-valued functions on the space of maximal ideals of $\mathcal{A}$ is a homomorphism of Banach algebras with norm less than or equal to 1 (see Appendix B). Since each $\sigma_{\mathbf{n}}$ is an isometry on S,

$$|\sigma_{\mathbf{n}}^*(\omega)| = 1 \text{ for all } \omega \in M(\mathcal{A}), \mathbf{n} \in \mathbb{Z}^n.$$

On the other hand, for any $z \in S$

$$\sum c_{\mathbf{n}} z_{\mathbf{n}+\mathbf{k}} = \left(\sum c_{\mathbf{n}} \sigma_{\mathbf{n}} z\right)_{\mathbf{k}},$$

so

$$S = \{z \in \ell_\infty : \left(\sum c_{\mathbf{n}} \sigma_{\mathbf{n}} z\right)(z) = 0\}.$$

It follows that for any $\omega \in M(\mathcal{A})$, $(\alpha_1, \ldots, \alpha_n) = (\sigma_{\mathbf{e}_1}^*(\omega), \ldots, \sigma_{\mathbf{e}_n}^*(\omega))$ is an n-tuple of complex numbers with modulus 1 annihilated by F. This contradicts the assumption on F and shows that B is an expansive neighbourhood for F.

4.2. If T_F is expansive, then it has no zero with all components on the unit circle K, so there can certainly not be a vector of unit roots at which F vanishes. It follows that (4.6) does not vanish for any period Λ, so there are only finitely many periodic points for each period.

The converse does not hold even when $n = 1$: there are integer polynomials with unit modulus zeros that are not unit roots (cf. Exercise 1.1).

4.3. Choose a sequence of periods $\Lambda_{r(j)} = (r(j)\mathbb{Z})^n$ with the property that there are finitely many points of period $\Lambda_{r(j)}$ for each j, and $r(j) \to \infty$ as $j \to \infty$. By (4.6) we have

$$|\mathrm{Per}_{\Lambda_{r(j)}}(T_F)| = \prod_{\mathbf{z} \in \Omega(\Lambda_{r(j)})} |F(\mathbf{z})| \neq 0,$$

and so

$$\frac{1}{r(j)^n} \log \prod_{\mathbf{z} \in \Omega(\Lambda_{r(j)})} |F(\mathbf{z})|$$

is a sequence of Riemann approximations to $m(F)$. In order to show these converge, notice that the only singularities of the integrand occur at points $(\alpha_1, \ldots, \alpha_n)$ with all coordinates equal to unit roots. It follows that

$$\rho\left((\alpha_1, \ldots, \alpha_n), \mathbf{z}\right) \gg \frac{1}{r(j)^2}$$

for all $\mathbf{z} \in \Omega(\Lambda_{r(j)})$, so the sequence converges to $m(F)$.

4.4. This is the two-dimensional case of a general result due to Hermite, stated in Lind [Lin96, Theorem 4.1]. A detailed proof is in Macduffee [Mac33, Theorem 22.2] in a slightly different setting. Let $A_i = \begin{bmatrix} a_i & b_i \\ 0 & c_i \end{bmatrix}$ for $i = 1, 2$ and assume that $\Lambda(A_1) = \Lambda(A_2)$.

This means there is a matrix $Q \in GL_2(\mathbb{Z})$ with $A_1 Q = A_2$. Multiplying out gives

$$\begin{bmatrix} q_{11}a_1 + q_{21}b_1 & q_{12}a_1 + q_{22}b_1 \\ q_{21}c_1 & q_{22}c_2 \end{bmatrix} = \begin{bmatrix} a_2 & b_2 \\ 0 & c_2 \end{bmatrix},$$

so $q_{21} = 0$ (since $a_1 \geq 1$). Also $q_{11}a_1 = a_2, q_{22}c_1 = c_2$, so $q_{11}q_{22} = 1$ since $a_1 c_1 = a_2 c_2 = |\mathbb{Z}^2/\Lambda(A_i)|$. So $q_{11} = q_{22} = \pm 1$. Since $c_1, c_2 \geq 1$ they are both 1. Then the identity reduces to

$$\begin{bmatrix} a_1 & q_{12}a_1 + b_1 \\ 0 & c_1 \end{bmatrix} = \begin{bmatrix} a_2 & b_2 \\ 0 & c_2 \end{bmatrix},$$

which shows that $q_{12} = 0$ since $0 \leq b_1 = b_2 \leq a_1 - 1 = a_2 - 1$. Thus Q is the identity.

We have shown that $A \mapsto \Lambda(A)$ is injective on the stated family of matrices.

To show that it is surjective, let Λ be a subgroup of $\mathbb{Z}^2$ with finite index. Then

$$\Lambda = \left\{ \begin{bmatrix} a & b \\ c & d \end{bmatrix} \begin{bmatrix} x \\ y \end{bmatrix} : (x, y) \in \mathbb{Z}^2 \right\}$$

for some non-singular matrix $\begin{bmatrix} a & b \\ c & d \end{bmatrix}$. We may apply an element $Q \in GL_2(\mathbb{Z})$ with the property that $Q \begin{bmatrix} a & b \\ c & d \end{bmatrix} = \begin{bmatrix} a' & b' \\ 0 & d' \end{bmatrix}$, so the subgroup is generated by $(a', 0)$ and (b', d'). Now add an integer multiple of the first generator to the second to ensure that b' lies between 0 and $a' - 1$ (this clearly does not change the subgroup) to see that Λ is of the required form.

4.5[1]. We have $|\mathrm{Per}_\Lambda(T)| = 1$ for all Λ, so (4.7) reduces to

$$\zeta_T(z) = \exp\left(\sum_{a=1}^{\infty} \sum_{c=1}^{\infty} \sum_{b=0}^{a-1} \frac{z^{ac}}{ac}\right)$$

$$= \exp\left(\sum_{a=1}^{\infty}\sum_{c=1}^{\infty}\frac{(z^a)^c}{c}\right)$$

$$= \exp\left(\sum_{a=1}^{\infty}-\log(1-z^a)\right)$$

$$= \prod_{a=1}^{\infty}(1-z^a)^{-1}.$$

If we write $\zeta_T(z) = \sum_{n=1}^{\infty}p(n)z^n$, then (assuming the product converges) $p(n)$ is the number of additive partitions of n. That is, the dynamical zeta function is the generating function for the partition function (see Hardy and Wright [HW79, Section 19.3]), which is known to be analytic for $|z| < 1$ and to have $|z| = 1$ as the natural boundary.

4.5[2]. As pointed out in Example 4.1[2], this is simply the shift action on the space $\prod_{\mathbb{Z}^2}\{0, 1, \ldots, |k|-1\}$. Thus $|\text{Per}_\Lambda(T_k)| = |k|^{|\mathbb{Z}^2/\Lambda|}$ for all periods Λ. Using $|k|z$ for z in the calculation in Exercise 4.5[1] shows that

$$\zeta_{T_k}(z) = \prod_{a=1}^{\infty}(1-|k|z)^{-1} = \sum_{n=1}^{\infty}p(n)(|k|z)^n.$$

It follows that the zeta function is analytic for $|z| < \exp(-h(T_k))$ and has $|z| = \exp(-h(T_k))$ as natural boundary. This is similar to the situation for single transformations in Exercise 2.7[4]. The next exercise shows that the radius of convergence will in general be strictly smaller than $\exp(-h(T_k))$.

4.6[1]. Following Example 4.3[1], write

$$B_\epsilon = \{z \in X : z_j = 0 \text{ for } |j| \leq R(\epsilon)\}$$

for some $R(\epsilon) \to \infty$ as $\epsilon \to 0$. Then

$$\bigcap_{0\leq k_i<m-1; i=1,2} T_{(-k_1,-k_2)}(B_\epsilon) = \{z \in X : z_j = 0 \text{ if } -R(\epsilon) \leq j \leq R(\epsilon)+m-1\}$$

so

$$\lambda\left(\bigcap_{0\leq k_i<m; i=1,2} T_{(-k_1,-k_2)}(B_\epsilon)\right) = \left(\frac{1}{|k|}\right)^{(2R(\epsilon)+m)}$$

giving

$$h(T) = 0.$$

4.6[2]. Let $A = \begin{bmatrix} a & b \\ 0 & c \end{bmatrix}$ with $0 \leq b \leq a-1$, $a, c \geq 1$. The subgroup $\Lambda(A)$ is generated by $(a, 0)$ and (b, c). Since $T_{(a,0)} = \sigma^0$, the identity, and $T_{(b,c)} = \sigma^c$, the points in $\text{Per}_{A(A)}(T)$ are exactly the points of period c under the shift σ on X. It follows that

$$|\text{Per}_{A(A)}(T)| = |k|^c.$$

Then

$$\zeta_T(z) = \exp\left(\sum_{a=1}^{\infty}\sum_{c=1}^{\infty}\sum_{b=0}^{a-1}\frac{|k|^c z^{ac}}{ac}\right)$$

$$= \exp\left(\sum_{a=1}^{\infty}\sum_{c=1}^{\infty}\frac{(|k|z^a)^c}{c}\right)$$

$$= \exp\left(\sum_{a=1}^{\infty}-\log(1-|k|z^a)\right)$$

$$= \frac{1}{(1-|k|z)(1-|k|z^2)(1-|k|z^3)\ldots}.$$

The smallest real pole of ζ_T is at $z = \frac{1}{|k|}$, which is strictly smaller than $\exp(-h(T))$ by Exercise 4.5[3]. There are infinitely many poles located at

$$\frac{1}{\sqrt[n]{|k|}}e^{2\pi ij/n} \text{ for } n \geq 1, 0 \leq j \leq n-1.$$

Thus there are infinitely many simple poles inside the disc $|z| < 1$ that cluster onto the unit circle. This means that ζ_T is meromorphic inside the disc and has the circle $|z| = 1$ as natural boundary (notice that the natural boundary is at $\exp(-h(T))$).

5.1[1]. Use the fact that $\alpha_1 + \alpha_2 + \alpha_3 = -a$, $\alpha_1\alpha_2 + \alpha_1\alpha_3 + \alpha_2\alpha_3 = 0$ and $\alpha_1\alpha_2\alpha_3 = -b$.

5.1[2]. This is fairly involved: see Silverman [Sil86, Appendix A] for the details.

5.2. See Silverman [Sil86, Section III.1] for the details.

5.3. Eliminate one of the variables.

5.4. Any rational solution gives rise to an integer triple a, b, c with $a^4 + b^2 = c^4$. A standard argument using Pythagorean triples shows that one of a, b or c is zero. Working back shows that the only rational point on E has $x = 0, \pm 1$ or is the point at infinity.

5.5. This is a restatement of the definition of addition on the curve.

5.6. Using the formulæ $2P = (-5, -16)$, $3P = (11, -32)$, $4P = (11, 32)$. Hence $3P = -4P$ giving $7P = 0$ so the order of P is 7. Notice also that $5P = (-5, 16)$, confirming that $2P = -5P$.

5.7. First notice that $-P = (x, -y)$; $2P = 0$ if and only if $P = -P$, that is $y = -y$.

5.8. From Example 5.14,

$$x(2P) = \frac{(x^2 + n^2)^2}{4(x^3 - n^2 x)}.$$

Now $3P = 0$ if and only if $2P = -P$, and $x(2P) = x(P)$ gives the quartic equation $3x^4 - 6x^2 n^2 + n^4 = 0$ which is quadratic in x^2. One pair of roots have $x^2 < 0$ so do not yield real points. Of the other two roots, only one gives $y^2 > 0$ so there are two real points (cf. Figure F.1).

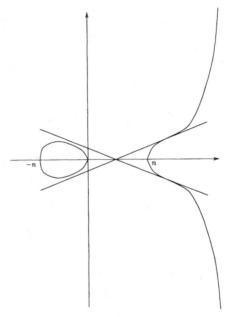

Fig. F.1. Real points on $y^2 = x^3 - n^2 x$

5.9. A point of order 4 must double to a point of order 2, but only one point of order 2 can be 'halved' to give a real point, as illustrated in Figure F.2.

The points marked P_1, P_2, P_3 and P_4 all double to Q, a point of order 2. Now let R denote any point of order 2 which is not Q. Then our group is $\{iR + jP_1 : i = 0, 1; j = 0, \ldots, 3\} \cong C_2 \times C_4$.

5.10. Factorize $x^3 - 1$ in $\mathbb{Z}[\omega]$ or $y^2 + 1$ in $\mathbb{Z}[i]$, then argue as in Theorem 5.8.

5.11. Calculate $x(2P)$ for $P = (x, y)$ and show that any cancellation between numerator and denominator is uniformly bounded.

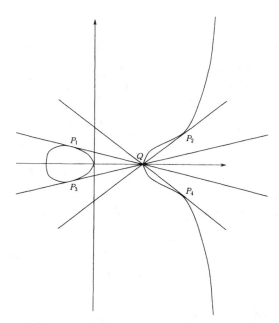

Fig. F.2. Points with order dividing four

5.12. The only zeros of the equations $(M^2 + n^2 N^2)^2 = N^2(M^2 - n^2 N^2) = 0$ are $M = N = 0$. For the curve $y^2 = x^3 + A$ assume that $A \in \mathbb{Z}$ for simplicity. Then

$$x(2P) = \frac{M^4 - 8MN^3 A}{4N(M^3 + AN^3)},$$

and the only zeros of the equations

$$0 = M^4 - 8MN^3 A = N(M^3 + AN^3)$$

are $M = N = 0$.

5.13. After two steps we see that the greatest common divisor of f and g must divide $(4a^3 + 27b^2)x$.

5.14. We already know by Exercise 5.7 that $y(Q) = 0$ corresponds to points of order 2. Assume that $y(Q) \neq 0$, and so $2Q \neq 0$. Writing $x = x(Q)$, $y = y(Q)$ we have from the proof of Theorem 5.13

$$x(2Q) = \frac{x^4 - 2x^2 a - 8xb + a^2}{4(x^3 + ax + b)}.$$

We know that $x(2Q) \in \mathbb{Z}$, so it follows that

$$4(x^3 + ax + b)|x^4 - 2x^2 a - 8xb + a^2.$$

Using the previous exercise, it follows that

$$y^2 = x^3 + ax + b|4a^3 + 27b^2.$$

5.15. If p is not prime, then [2] and [3] fail. For example, with $p = 10$ we have $|2|_{10} = 1$, $|5|_{10} = 1$ while $|10|_{10} = \frac{1}{10}$ (showing [2] does not hold), and $|\frac{1}{2}|_{10} = |\frac{1}{5}|_{10} = 1$ while $|\frac{1}{2} + \frac{1}{5}|_{10} = 10$.

5.16. This follows immediately from unique factorization in the integers.

6.1[1]. It is sufficient to prove that $\gamma = \sum'_{(m,n)} \frac{1}{(m+ni)^6} = 0$. After rearranging terms $-\gamma = \frac{1}{i^6}\gamma = \gamma$. Clearly $a \in \mathbb{R}$ since every term appears with its complex conjugate.

6.1[2]. This is similar: if $\theta = \sum'_{(m,n)} \frac{1}{(m+ni)^4}$ then $\omega\theta = \theta$ after re-arranging terms. To see that $b \in \mathbb{R}$ argue as in [1].

6.2. Notice that $E(\mathbb{C}) \cong \mathbb{C}/L$ as groups. It follows that the points with order dividing n are all those of the form $(r_1\omega_1 + r_2\omega_2)/n$, $0 \le r_1, r_2 < n$.

6.3. What are the points of order 2 on an elliptic curve according to Exercise 5.7?

Alternative solution: by periodicity,

$$\wp'_L(\omega_1/2) = \wp'_L(\omega_1/2 - \omega_1) = \wp'_L(-\omega_1/2).$$

Since $\wp'_L$ is an odd function, it follows that $\wp'_L(-\omega_1/2) = -\wp'_L(\omega_1/2)$, so $\wp'_L(\omega_1/2) = 0$. A similar argument applies to the other two points.

6.4[1]. Find the coefficient of z^2 in $g(z)$ in the proof of Theorem 6.4[1] in two ways. First in terms of G_4 and G_8 and secondly notice that this must be zero since $g(z)$ is identically zero.

6.4[2]. Use $g(z)$ again and induction.

6.5[1]. The map $z + L \mapsto cz + L'$ gives an isomorphism between $\mathbb{C}/L$ and $\mathbb{C}/L'$. For the second part, see Silverman [Sil86, Chapter VI, Section 4], or one may argue as follows. Note that E and E' are given by quotients $\mathbb{C}/L$ and $\mathbb{C}/L'$. An isomorphism $f : E \to E'$ between them is an isomorphism of complex Lie groups, so in a neighbourhood of $z = 0$ the isomorphism is given by a convergent power series,

$$f(z) = a_0 + a_1 z + \dots.$$

On the other hand, $f(z+z') - f(z) - f(z') \in L'$ so for $|z|, |z'|$ small, $f(z+z') = f(z) + f(z')$. The only convergent power series with this property is $f(z) = \alpha z$ for some $\alpha \in \mathbb{C}$. Thus, in a neighbourhood of $z = 0$, f has the desired form. Given any $z \in \mathbb{C}/L$, for a large enough N, $\frac{z}{N}$ is close enough to 0 to have $f(\frac{z}{N}) = \alpha\frac{z}{N}$, so $f(z) = \alpha z$ for all z.

6.5[2]. Using Theorem 6.4, we see that a is a multiple of $\sum_{\ell \neq 0} \ell^{-4}$ while b is a multiple of $\sum_{\ell \neq 0} \ell^{-6}$. So, under the isomorphism $a \mapsto c^{-4}a = a'$, $b \mapsto c^{-6}b = b'$. Thus, the point $(x', y') = (c^{-2}x, c^{-3}y)$ lies on the curve

$$y'^2 = x'^3 + a'x' + b'$$

and conversely.

6.6. This formula is a special case of a very general result applicable to any holomorphic function f with simple zeros $a_1, a_2, \ldots$ with $|a_n| \to \infty$ (see Whittaker and Watson [WW63, Section 7.5] for more on this, but we would not recommend using this source for general background in complex analysis – see Halmos [Hal88, Section 22]). The derivative f' is holomorphic for all z, so $g(z) = \frac{f'(z)}{f(z)}$ can only have singularities at the points $a_1, a_2, \ldots$. The Taylor expansion at a_r

$$f(z) = (z - a_r)f'(a_r) + \frac{(z - a_r)^2}{2}f''(a_r) + \ldots$$

shows that

$$f'(z) = f'(a_r) + (z - a_r)f''(a_r) + \ldots$$

so, at each point a_r, g has a simple pole with residue $+1$.

Now assume that there is a sequence of circular contours Γ_m, with radii $r_m \to \infty$ such that

$$|g(z)| \leq M \text{ for all } z \in \Gamma_m \text{ and } m = 1, 2, \ldots. \tag{F.10}$$

Fix w not equal to any a_r. By Cauchy's theorem,

$$\frac{1}{2\pi i}\int_{\Gamma_m} \frac{g(z)}{z - \omega}dz = g(\omega) + \sum_r \frac{1}{a_r - \omega},$$

where the sum runs over those r for which a_r lies inside Γ_m. On the other hand,

$$\frac{1}{2\pi i}\int_{\Gamma_m} \frac{g(z)}{z - \omega}dz = \frac{1}{2\pi i}\int_{\Gamma_m} \frac{g(z)}{z}dz + \frac{\omega}{2\pi i}\int_{\Gamma_m} \frac{g(z)}{z(z - \omega)}dz$$

$$= g(0) + \sum_r \frac{1}{a_r} + \frac{\omega}{2\pi i}\int_{\Gamma_m} \frac{g(z)}{z(z - x)}dz$$

if g is analytic at the origin. As $m \to \infty$, the assumption (F.10) shows that the last integral converges to 0 as $m \to \infty$. It follows that

$$g(\omega) = g(0) + \sum_{r=1}^{\infty}\left(\frac{1}{a_r - \omega} - \frac{1}{a_r}\right).$$

If $|a_n| \le |a_{n+1}|$ for all n, then the series is uniformly convergent on any disc (by grouping terms corresponding to poles of given modulus), so the series may be integrated term-by-term. Doing this and exponentiating gives

$$g(z) = Ce^{g(0)z} \prod_{r=1}^{\infty} \left\{ \left(1 - \frac{z}{a_r}\right) e^{z/a_r} \right\} \tag{F.11}$$

for some constant C independent of z. Taking $z = 0$ shows that $C = g(0)$.

Now apply this to $f(z) = \frac{\sin z}{z}$ (defined as written for $z \ne 0$ and extended to all z by defining $f(0) = 1$). Then $g(z) = \frac{f'(z)}{f(z)}$ has singularities at $r\pi$ for $r = \pm 1, \pm 2, \dots$ and meets the criteria in the discussion above. It follows by (F.11) that

$$\frac{\sin z}{z} = \prod_{n=1}^{\infty} \left\{ \left(1 - \frac{z}{n\pi}\right) e^{z/n\pi} \right\} \left\{ \left(1 + \frac{z}{n\pi}\right) e^{-z/n\pi} \right\} = \prod_{n=1}^{\infty} \left(1 - \frac{z^2}{n^2\pi^2}\right).$$

6.7. Use the triangle inequality.

6.8[1]. The function on the right has a triple pole at each lattice point and a simple zero at each point of the form $\frac{1}{2}\ell$, $\ell \in L$. It follows that $\frac{\wp_L'(z)\sigma_L^4(z)}{\sigma_L(2z)}$ is a constant. Let $z \to 0$ and recall that $\wp_L'(z) = \frac{-2}{z^3} + \dots$, $\sigma_L^4(z) = z^4 + \dots$ and $\sigma_L(2z) = 2z + \dots$ to show that this constant value is -1. Alternatively, let $z \to a$ in the formula in Lemma 6.9.

Similarly, to obtain the formula for $\lambda_L(2z)$, let $z \to a$ in Theorem 6.11[1].

6.8[2]. Write $[e_1, e_2, e_3] = [\wp_L(\omega_1/2), \wp_L(\omega_2/2), \wp_L((\omega_1 + \omega_2)/2)]$. Then

$$|\wp_L'(z)|^2 = 4\,|(\wp_L(z) - e_1)\,(\wp_L(z) - e_2)\,(\wp_L(z) - e_3)|$$

so

$$2 \int_{\Pi} \log |\wp_L'(z)| d\mu_{\Pi} = \log 4 + 2 \sum_{i=1}^{3} \lambda_L(e_i).$$

On the other hand, the parallelogram law gives

$$\log |e_i - e_j| + 2\lambda_L(e_k) = 2\lambda_L(e_i) + 2\lambda_L(e_j)$$

for distinct i, j, k with $\{i, j, k\} = \{1, 2, 3\}$. Adding the three equations gives

$$\tfrac{1}{2} \log |\Delta| = 2 \sum_{i=1}^{3} \lambda_L(e_i)$$

using the equation

$$|\Delta|^2 = |(e_1 - e_2)(e_1 - e_3)(e_2 - e_3)|.$$

Therefore

$$\int_{\Pi} \log |\wp'_L(z)| = \log 2 + \frac{1}{4} \log |\Delta|.$$

6.9[1]. Use the parallelogram law to write down three linear equations for λ_1, λ_2 and λ_3.

6.9[2]. Use the parallelogram law and the duplication formula.

6.10[1]. It is enough to measure the size of the set on which $|\wp_L(z) - \wp_L(a)|$ is less than δ. Use Lemma 6.9 and the fact that $|\sigma_L(z)/z|$ is uniformly bounded for $z \notin L$. Note that special care needs to be taken if $2a \in L$ (that is, if $\wp'_L(a) = 0$).

6.10[2]. Use [1] and induction (as in Lemma 3.8).

6.10[3]. Note that Nz is really $Nz \bmod L$. Thus $Nz = a$ has N^2 solutions in Π. A δ-neighbourhood of any solution of $\wp_L(Nz) - \wp_L(a) = 0$ has measure $c\delta^{1/2}/N^2$ and there are N^2 such points.

6.10[4]. Compare with Exercise 1.6. Change the variable $\omega = Nz$ and use the periodicity of $\wp_L$.

6.10[5]. Use the Weierstrass approximation Theorem B.2 to write the integrand as a sum of functions periodic on Π.

6.11. If n_1 divides n_2 then $\psi^2_{n_1}$ divides $\psi^2_{n_2}$ in $\mathbb{Q}[x]$ since $n_1\rho = 0$ implies $n_2\rho = 0$. Now the content of $\psi^2_{n_1}$ divides the content of $\psi^2_{n_2}$ by the remark at the end of Appendix C, so $\psi^2_{n_1}$ divides $\psi^2_{n_2}$ in $\mathbb{Z}[x]$ also. The claim follows at once.

6.12[1]. This can be done with the usual method of comparing zeros and poles then normalizing by a constant.

6.12[2]. This is in Silverman [Sil94, Exercise 6.4, Chapter VI], where he quotes Whittaker and Watson [WW63, misc. Exercise 33, Chapter XX]. The approach there is to first prove by induction that

$$\det \begin{bmatrix} 1 & \wp_L(z_0) & \wp'_L(z_0) & \cdots & \wp_L^{(n-1)}(z_0) \\ 1 & \wp_L(z_1) & \wp'_L(z_1) & \cdots & \wp_L^{(n-1)}(z_1) \\ \vdots & & & & \vdots \\ 1 & \wp_L(z_n) & \wp'_L(z_n) & \cdots & \wp_L^{(n-1)}(z_n) \end{bmatrix}$$

$$= (-1)^{n(n-1)/2} 1!2! \ldots n! \frac{\sigma_L(z_0 + z_1 + \ldots + z_n) \prod_{0 \le \lambda < \mu \le n} \sigma_L(z_\lambda - z_\mu)}{\sigma_L^{n+1}(z_0) \ldots \sigma_L^{n+1}(z_n)},$$

a result due to Frobenius and Stickelberger.

6.13. By the elliptic Jensen formula,

$$m_E(F') \leq m_E(F) + O(d) \text{ and } m_E(F) = m(F) + O(d)$$

so the result follows from Exercise 1.6[2].

6.14. If $q < 0$ then $E(\mathbb{R}) \cong \mathbb{R}/\mathbb{Z}$ so $E_n(\mathbb{R}) = C_n$ clearly. The case of odd n is similar. If $q > 0$ and n is even then the elements of order n in $C_2 \times \mathbb{R}/\mathbb{Z}$ are given by any element in C_2 in the first coordinate and any element of C_n in the second.

F.2 List of Questions

For convenience, the 14 open problems raised in the text are collected here.

1. Can you find an elementary proof of (1.20), which says

$$\int_0^1 \log \left| \sum_{j=1}^d \frac{1}{e^{2\pi i \theta} - \alpha_j} \right| d\theta \leq \log d.$$

for any complex numbers $\alpha_1, \ldots, \alpha_d$? [Section 1.2]

2. Is there a meaningful *lower* bound for $m(F')$? Any lower bound for $m(F')$ in terms of $m(F)$ must necessarily involve some dependence on the constant coefficient $F(0)$, as the example $F(x) = x - N$ shows. Experiment to find a sharp lower bound for $m(F')$. As a starting point, notice that $F(x) = x - N$ satisfies

$$m(F) + \log |d/F(0)| \leq m(F').$$

[Section 1.2]

3. Let $F(x) = x^3 - x - 1$. Are there infinitely many primes in the sequence $\Delta_n(F)$? [Section 1.6]

4. Can the arithmetic properties of the sequences considered by Lehmer be developed in the same way that the arithmetic properties of binary sequences have? See Ribenboim [Rib95a], Stewart [Ste77] for background, and van der Poorten [Poo89] and references therein for an introduction to the large body of results on recurrence sequences in general. [Section 1.6]

5. Define the *dimension* of $F \in \bigcup_{n \geq 1} \mathbb{Z}[x_1, \ldots, x_n]$ to be $\dim(F) = \dim(\mathcal{C}(F))$, the dimension of the Newton polygon of F. Theorem 3.10 shows that an *irreducible* polynomial with $m(F) = 0$ must be one-dimensional. Define

$$\lambda(m) = \min\{m(F) : F \text{ is irreducible and } \dim(F) = m\}.$$

Is it true that $\lambda(m) \to \infty$ as $m \to \infty$? (cf. Boyd, [Boy81, p. 461]). [Section 3.3]

6. Let $F \in \mathbb{C}[x, y]$ be a polynomial in which both variables appear, and assume that $F(z, z^d) = 0$ has a solution on K. Is there a rational $q > 0$ (depending on F) for which

$$\frac{C_1}{k} \epsilon^q < \mu \left(\{ z \in K : |F(z, z^d)| < \epsilon \} \right) < C_2 k \epsilon^q,$$

where k is the number of non-zero coefficients in $F(z, z^d)$ and C_1, C_2 are absolute constants? [Section 3.5]

7. Can Exercise 3.13 be used to give an alternative proof of Theorem 3.10? All the ideas needed are in the paper of Dobrowolski, Lawton and Schinzel [DLS83]. [Section 3.5]

8. If $F \in \mathbb{Z}[x_1, \ldots, x_n]$ has the property that T_F has finitely many points of each period but T_F is not expansive, does it follow that the growth rate of periodic points exists? [Section 4.3]

9. Is there an analogue of Theorem 2.16[2] for $\mathbb{Z}^n$-actions? [Section 4.3]

10. Is it true that the dynamical zeta function of T_F is meromorphic in the disc $|z| < e^{-h(T_F)}$ and has the circle $|z| = e^{-h(T_F)}$ as natural boundary? (cf. Lind [Lin96, Conjecture 7.1]). [Section 4.4]

11. Suppose E denotes an arbitrary elliptic curve. Do there exist numbers $\alpha_1, \ldots, \alpha_t$ in $\mathbb{R}$ with the following property: for any $F \in \mathbb{Z}[x]$ with degree d having real roots,
$$m_E(F) > cd, \qquad c = c(E) > 0,$$
provided $F(\alpha_1) \ldots F(\alpha_t) \neq 0$? Clearly some auxiliary conditions on the zeros of F will necessary, just as in Schinzel's theorem. [Section 6.2]

12. Can you improve the bound to one of the form

$$m_E(F') \leq m_E(F) + c_2 \log d$$

for some $c_2 = c_2(E) > 0$? [Section 6.5]

13. What is the analogue of Corollary 6.25 when q is positive? [Section 6.6]

14. Can you construct a canonical family of maps (parametrized by elliptic curves defined over the rationals and integer polynomials) whose periodic points are related to the formula for $E_n(F)$ in Section 6.4 and whose topological entropies coincide with the elliptic Mahler measures of integer polynomials? [Section 6.6]

G. List of Notation

The symbols $\mathbb{N}, \mathbb{Z}, \mathbb{Q}, \mathbb{R}, \mathbb{C}$ denote the natural numbers, integers, rationals, reals and complex numbers respectively. The additive circle $\mathbb{R}/\mathbb{Z}$ is denoted $\mathbb{T}$, and the multiplicative circle $\{z \in \mathbb{C} : |z| = 1\}$ is denoted K. The notation $f(n) = O(g(n))$ is used to mean that there exists a constant A independent of n for which $f(n) \leq Ag(n)$. The relation $f(n) = O(g(n))$ will also be written $f \ll g$ when it is being used to express the fact that two functions are comparable: $f \ll g \ll f$. The notation $f(x) \sim g(x)$ is used for small values of x to mean that $f(x)/g(x) \to 1$ as $x \to 0$.

Bibliography

[Abr59]   L.M. Abramov, *The entropy of an automorphism of a solenoidal group*, Teor. Verojatnost. i Primenen. **4** (1959), 249–254 (Russian), English transl. Theory of Prob. and Applic. **4** (1959), 231–236.

[AKM65]   R. Adler, A. Konheim, and M. McAndrew, *Topological entropy*, Transactions of the Amer. Math. Soc. **114** (1965), 309–319.

[AM65]   E. Artin and B. Mazur, *On periodic points*, Annals of Math. **81** (1965), 82–99.

[Amo96]   F. Amoroso, *Algebraic numbers close to 1 and variants of Mahler's measure*, Journal of Number Theory **60** (1996), 80–96.

[Bak66]   A. Baker, *Linear forms in the logarithms of algebraic numbers I*, Mathematika **13** (1966), 204–216.

[Bak67a]   A. Baker, *Linear forms in the logarithms of algebraic numbers II*, Mathematika **14** (1967), 102–107.

[Bak67b]   A. Baker, *Linear forms in the logarithms of algebraic numbers III*, Mathematika **14** (1967), 220–228.

[Bak68]   A. Baker, *Linear forms in the logarithms of algebraic numbers IV*, Mathematika **15** (1968), 204–216.

[Bak75]   A. Baker, *Transcendental Number Theory*, Cambridge University Press, Cambridge, 1975.

[BBM79]   M. Barnsley, D. Bessis, and P. Moussa, *The Diophantine moment problem and the analytic structure in the activity of the ferromagnetic Ising model*, Journal of Math. Phys. **20** (1979), 535–546.

[BE95]   P. Borwein and T. Erdélyi, *Polynomials and Polynomial Inequalities*, Springer, New York, 1995.

[Ber69]   K. Berg, *Convolution of invariant measures, maximal entropy*, Math. Systems Theory **3** (1969), 146–151.

[BLS75]   J. Brillhart, D.H. Lehmer, and J.L. Selfridge, *New primality criterion and factorizations of* $2^m \pm 1$, Math. Comp. **29** (1975), 620–647.

[BLS+83]   J. Brillhart, D.H. Lehmer, J.L. Selfridge, B. Tuckerman, and S.S. Wagstaff Jr., *Factorizations of* $b^n \pm 1, b = 2, 3, 5, 6, 7, 10, 11, 12$ *up to high powers*, American Math. Society, Providence, R.I., 1983.

[BM40]   G. Billing and K. Mahler, *On exceptional points on cubic curves*, Journal of the London Math. Soc. **15** (1940), 32–43.

[BM71]   P.E. Blanksby and H.L. Montgomery, *Algebraic integers near the unit circle*, Acta Arith. **18** (1971), 355–369.

[Bow71]   R. Bowen, *Entropy for group endomorphisms and homogeneous spaces*, Transactions of the Amer. Math. Soc. **153** (1971), 401–414.

[Bow78]   R. Bowen, *On Axiom A diffeomorphisms*, Amer. Math. Soc., Providence, 1978, CBMS Reg. Conf. Series 35.

[Boy80]   D.W. Boyd, *Reciprocal polynomials having small measure*, Math. Comp. **35** (1980), 1361–1377.

[Boy81]   D.W. Boyd, *Speculations concerning the range of Mahler's measure*, Canadian Math. Bulletin **24** (1981), 453–469.

[Boy89]   D.W. Boyd, *Reciprocal polynomials having small measure II*, Math. Comp. **53** (1989), 353–357, S1–S5.

[Boy92]   D.W. Boyd, *Two sharp inequalities for the norm of a factor of a polynomial*, Mathematika **39** (1992), 341–349.

[Boy98]   D.W. Boyd, *Mahler's measure and special values of L-functions*, Experimental Math. **7** (1998), 37–82.

[BP93]   R. Burton and R. Pemantle, *Local characteristics, entropy and limit theorems for spanning trees and domino tilings via transfer–impedance*, Annals of Probability **21** (1993), 1329–1371.

[Bru93]   J.W. Bruce, *A really trivial proof of the Lucas–Lehmer test*, American Math. Monthly **100** (1993), 370–371.

[BSW89]   P.T. Bateman, J.L. Selfridge, and S.S. Wagstaff Jr., *The new Mersenne conjecture*, American Math. Monthly **96** (1989), 125–128.

[Cas49]   J.W.S. Cassels, *A note on division values of $\wp(u)$*, Math. Proc. Camb. Phil. Soc. **45** (1949), 167–172.

[Cas86]   J.W.S. Cassels, *Local Fields*, Cambridge University Press, Cambridge, 1986.

[Cas91]   J.W.S. Cassels, *Lectures on Elliptic Curves*, Cambridge University Press, Cambridge, 1991.

[CEW97]   V. Chothi, G. Everest, and T. Ward, *S-integer dynamical systems: periodic points*, Journal für die Reine und. angew. Math. **489** (1997), 99–132.

[CKP83]   G.F. Carrier, M. Krook, and C.E. Pearson, *Functions of a Complex Variable*, Hod Books, Ithaca, NY, 1983.

[Coo98]   M.J. Cooker, *The accurate summation of some awkward series*, The Mathematical Gazette **82** (March 1998), 48–55.

[CS82]   D.C. Cantor and E.G. Strauss, *On a conjecture of D.H. Lehmer*, Acta Arith. **42** (1982), 97–100.

[Dav95]   S. David, *Minorations de formes linéaires de logarithmes elliptiques*, Mem. Soc. Math. France **62** (1995), 143pp.

[Deg97]   J. Degot, *Finite–dimensional Mahler measure of a polynomial and Szegö's theorem*, Journal of Number Theory **62** (1997), 422–427.

[DEM+98]   P. D'Ambros, G. Everest, R. Miles and T. Ward, *Dynamical systems arising from elliptic curves*, preprint (1998).

[Den97]   C. Deninger, *Deligne periods of mixed motives, K–theory and the entropy of certain Z^n-actions*, Journal of the Amer. Math. Soc. **10** (1997), 259–281.

[DLS83]   E. Dobrowolski, W. Lawton, and A. Schinzel, *On a problem of Lehmer*, Studies in Pure Mathematics (Boston, Mass.), Birkhaüser, 1983, pp. 135–144.

[Dob79]   E. Dobrowolski, *On a question of Lehmer and the number of irreducible factors of a polynomial*, Acta Arith. **34** (1979), 391–401.

[Dob81]   E. Dobrowolski, *On a question of Lehmer*, Mem. Soc. Math. France **No. 2** (1980-81), 35–39.

[Dob91]   E. Dobrowolski, *Mahler's measure of a polynomial in function of the number of its coefficients*, Canadian Math. Bulletin **34** (1991), 186–195.

[Dre98]   G.P. Dresden, *Orbits of algebraic numbers with low heights*, Math. Comp. **67** (1998), 815–820.

[Dur81]   A. Durand, *On Mahler's measure of a polynomial*, Proceedings of the Amer. Math. Soc. **83** (1981), 75–76.

[EF96]   G.R. Everest and Bríd ní Fhlathúin, *The elliptic Mahler measure*, Math. Proc. Camb. Phil. Soc. **120** (1996), 13–25.

[Ein]   M. Einsiedler, *A generalization of Mahler measure and its application in algebraic dynamical systems*, Acta Arithmetica (to appear).

[Eis66]   M. Eisenberg, *Expansive automorphisms of finite-dimensional spaces*, Fundamenta Math. **LIX** (1966), 307–312.

[EP]   G.R. Everest and C. Pinner, *Bounding the elliptic Mahler measure II*, Journal of the London Math. Soc. (to appear).

[Erd32]   P. Erdös, *Beweis eines Satzes von Tschebyschef*, Acta. Litt. Acad. Sci. (Szeged) **5** (1932), 194–198.

[Eve98]   G.R. Everest, *Measuring the height of a polynomial*, Math. Intelligencer **20** (1998), 9–16.

[Eve]   G.R. Everest, *The elliptic analogue of Jensen's formula*, Journal of the London Math. Soc. (to appear).

[EW97]   M. Einsiedler and T. Ward, *Fitting ideals for finitely presented algebraic dynamical systems*, preprint (1997).

[EW98]   G. Everest and T. Ward, *A dynamical interpretation of the global canonical height on an elliptic curve*, Experimental Math. **7** (1998), 305–316.

[Fal91]   G. Faltings, *Diophantine approximation on abelian varieties*, Annals of Math. **133** (1991), 549–576.

[Fis61]   M.E. Fisher, *Statistical mechanics of dimers on a plane lattice*, Phys. Rev. **124** (1961), 1664–1672.

[FL96]   L. Flatto and J.C. Lagarias, *The lap-counting function for linear mod one transformations I: explicit formulas and renormalizability*, Ergodic Theory and Dynamical Systems **16** (1996), 451–491.

[FL97a]   L. Flatto and J.C. Lagarias, *The lap-counting function for linear mod one transformations II: the Markov chain for generalized lap numbers*, Ergodic Theory and Dynamical Systems **17** (1997), 123–146.

[FL97b]   L. Flatto and J.C. Lagarias, *The lap-counting function for linear mod one transformations III: the period of a Markov chain*, Ergodic Theory and Dynamical Systems **17** (1997), 369–403.

[Fla97]   V. Flammang, *Inégalités sur la mesure de Mahler d'un polynôme*, Journal de Théorie des Nombres de Bordeaux **9** (1997), 69–74.

[FLP94]   L. Flatto, J.C. Lagarias, and B. Poonen, *The zeta function of the beta transformation*, Ergodic Theory and Dynamical Systems **14** (1994), 237–266.

[Gan59]   F.R. Gantmacher, *Matrix Theory*, vol. I & II, Chelsea Publishing Company, New York, 1959, Translated from the Russian by K. Hirsch.

[Gel60]   A.O. Gelfond, *Transcendental and Algebraic Numbers*, Dover, New York, 1960, Translated from the Russian by L.F. Boron.

[Gon50]   J.V. Gonçalves, *L'inégalité de W. Specht*, Faculdade de Ciências, Univ. de Lisboa (1950), 167–171.

[GR94]   I.S. Gradshteyn and I.M. Ryzhik, *Table of Integrals, Series and Products*, Academic Press, Boston, 1994.

[Hal43]   P.R. Halmos, *On automorphisms of compact groups*, Bulletin of the Amer. Math. Soc. **49** (1943), 619–624.

[Hal88]   P.R. Halmos, *Some books of auld lang syne*, A Century of Mathematics in America (P. Duren, ed.), vol. 1, 1988, pp. 131–174.

[Har77]   R. Hartshorne, *Algebraic Geometry*, Springer, New York, 1977.

[HLP34]   G. Hardy, J.E. Littlewood, and G. Polya, *Inequalities*, Cambridge University Press, Cambridge, 1934.

[Hof78]   F. Hofbauer, *β-shifts have unique maximal measures*, Monatshefte Math. **85** (1978), 189–198.

[HR63]   E. Hewitt and K. Ross, *Abstract Harmonic Analysis*, Springer, New York, 1963.

[HS88]   M. Hindry and J.H. Silverman, *The canonical height and integral points on elliptic curves*, Inventiones Math. **93** (1988), 419–450.

[HS93]   G. Hoehn and N.-P. Skoruppa, *Un resultat de Schinzel*, Journal de Théorie des Nombres de Bordeaux **5** (1993), 185.

[Hus87]   D. Husemöller, *Elliptic Curves*, Springer, New York, 1987.

[HW79]   G.H. Hardy and E.M. Wright, *An Introduction to the Theory of Numbers*, fifth ed., Clarendon Press, Oxford, 1979.

[Kas61]   P.W. Kasteleyn, *The statistics of dimers on a lattice*, Physica **27** (1961), 1209–1225.

[Kat76]   Y. Katznelson, *An Introduction to Harmonic Analysis*, Dover, New York, 1976.

[KH95]   A. Katok and B. Hasselblatt, *Introduction to the Modern Theory of Dynamical Systems*, Cambridge University Press, Cambridge, 1995.

[Knu81]   D.E. Knuth, *The Art of Computer Programming, Vol. 2: Seminumerical Algorithms*, 2 ed., Addison–Wesley, Reading, MA, 1981.

[Kob77]   N. Koblitz, *p–adic Numbers, p–adic Analysis, and Zeta Functions*, Springer, New York, 1977.

[Kob84]   N. Koblitz, *Introduction to Elliptic Curves and Modular Forms*, Springer, New York, 1984.

[KS89]   B. Kitchens and K. Schmidt, *Automorphisms of compact groups*, Ergodic Theory and Dynamical Systems **9** (1989), 691–735.

[Lan73]   S. Lang, *Real Analysis*, Addison-Wesley, New York, 1973.

[Lan78]   S. Lang, *Elliptic Curves: Diophantine Analysis*, Springer, New York, 1978.

[Lan88]   S. Lang, *Introduction to Arakelov Theory*, Springer, New York, 1988.

[Lau83]   M. Laurent, *Minoration de la hauteur de Néron-Tate*, Séminaire de Théorie de Nombres Paris 1981-2, Birkhaüser, Basel, 1983, pp. 137–152.

[Law75]   W. Lawton, *Heights of algebraic numbers and Szegö's theorem*, Proceedings of the Amer. Math. Soc. **49** (1975), 47–50.

[Law83]   W. Lawton, *A problem of Boyd concerning geometric means of polynomials*, Journal of Number Theory **16** (1983), 356–362.

[Leh30]   D.H. Lehmer, *An extended theory of Lucas' functions*, Annals of Math. **31** (1930), 419–448.

[Leh32]   D.H. Lehmer, *On Euler's totient function*, Bulletin of the Amer. Math. Soc. **38** (1932), 745–757.

[Leh33]   D.H. Lehmer, *Factorization of certain cyclotomic functions*, Annals of Math. **34** (1933), 461–479.

[Leh47]   D.H. Lehmer, *On the factors of $2^n \pm 1$*, Bulletin of the Amer. Math. Soc. **53** (1947), 164–167.

[Lie67]   E.H. Lieb, *Residual entropy of square ice*, Phys. Rev. **162** (1967), 162–172.

[Lin74]   D.A. Lind, *Ergodic automorphisms of the infinite torus are Bernoulli*, Israel Journal of Math. **17** (1974), 162–168.

[Lin77]   D.A. Lind, *The structure of skew products with ergodic group automorphisms*, Israel Journal of Math. **28** (1977), 205–248.

[Lin82]   D.A. Lind, *Dynamical properties of quasihyperbolic toral automorphisms*, Ergodic Theory and Dynamical Systems **2** (1982), 49–68.

[Lin96]   D.A. Lind, *A zeta function for Z^d-actions*, Ergodic Theory of Z^d-actions (M. Pollicott and K. Schmidt, eds.), Cambridge University Press, Cambridge, 1996, pp. 433–450.

[Lou83]   R. Louboutin, *Sur la mesure de Mahler d'un nombre algebrique*, Comptes Rendus Acad. Sci. Paris **296** (1983), 707–708.

[LSW90]   D.A. Lind, K. Schmidt, and T. Ward, *Mahler measure and entropy for commuting automorphisms of compact groups*, Inventiones Math. **101** (1990), 593–629.

[Lub97]   D.S. Lubinsky, *Small values of polynomials: Cartan, Pólya and others*, J. of Inequal. and Appl. **1** (1997), 199–222.

[Luc79]   F. Lucas, *Géométrie des polynômes*, J. École Polytech. **46** (1879), 1–33.

[Lut37]   E. Lutz, *Sur l'equation $y^2 = x^3 - Ax - B$ dans les corps p–adic*, Journal für die Reine und. angew. Math. **177** (1937), 237–247.

[LW88]   D.A. Lind and T. Ward, *Automorphisms of solenoids and p-adic entropy*, Ergodic Theory and Dynamical Systems **8** (1988), 411–419.

[Mac33]   C.C. Macduffee, *The Theory of Matrices*, Verlag von Julius Springer, Berlin, 1933.

[Mah60]   K. Mahler, *An application of Jensen's formula to polynomials*, Mathematika **7** (1960), 98–100.

[Mah61]   K. Mahler, *On the zeros of the derivative of a polynomial*, Proceedings of the Royal Society **264** (1961), 145–154.

[Mah62]   K. Mahler, *On some inequalities for polynomials in several variables*, Journal of the London Math. Soc. **37** (1962), 341–344.

[Mah64]   K. Mahler, *An inequality for the discriminant of a polynomial*, Michigan Math. Journal **11** (1964), 257–262.

[Mai97]   V. Maillot, *Géométrie d'Arakelov des variétés toriques et fibrés en droite intégrables*, preprint (1997).

[Mar49]   M. Marden, *Geometry of Polynomials*, American Mathematical Society, Providence, RI, 1949.

[Mas89]   D.W. Masser, *Counting points of small height on elliptic curves*, Bull. Soc. Math. France **117** (1989), 247–265.

[MGB84]   P. Moussa, J.S. Geronimo, and D. Bessis, *Ensembles de Julia et propriétés de localisation des familles itérées d'entiers algébriques*, Comptes Rendus Acad. Sci. Paris **299** (1984), 281–284.

[Mil86]   J.S. Milne, *Abelian varieties*, Arithmetic Geometry (G. Cornell and J.H. Silverman, eds.), Springer, New York, 1986, pp. 103–150.

[Mor90]   F. Morain, *Distributed primality proving and the primality of $(2^{3539} + 1)/3$*, Proceedings EUROCRYPT 1990, Springer, New York, 1991.

[Mor22]   L.J. Mordell, *On the rational solutions of the indeterminate equations of the third and fourth degrees*, Proc. Cambridge Phil. Soc. **21** (1922), 179–192.

[Mos95]   M.J. Mossinghoff, *Algorithms for the determination of polynomials with small Mahler measure*, Ph.D. thesis, The University of Texas at Austin, 1995.

[Mos98]   M.J. Mossinghoff, *Polynomials with small Mahler measure*, Math. Comp. **67** (1998), 1697–1705.

[Mou83]   P. Moussa, *Problème diophantien des moments et modèle d'Ising*, Annales de l'Institut Henri Poincaré **38** (1983), 309–347.

[Mou86]   P. Moussa, *Diophantine properties of Julia sets*, Chaotic Dynamics and Fractals (M.F. Barnsley and S. Demko, eds.), Academic Press, 1986, pp. 215–227.

[Mou90]   P. Moussa, *The Ising model and the Diophantine moment problem*, Number Theory and Physics (J.-M. Luck, P. Moussa, and M. Waldschmidt, eds.), Springer, 1990, pp. 264–275.

[MPV98]   M.J. Mossinghoff, C.G. Pinner, and J.D. Vaaler, *Perturbing polynomials with all their roots on the unit circle*, Math. Comp. **67** (1998), 1707–1726.

[MW98]   G. Morris and T. Ward, *Entropy bounds for endomorphisms commuting with K actions*, Israel Journal of Math. **106** (1998), 1–12.

[Mye84]   G. Myerson, *A measure for polynomials in several variables*, Canadian Math. Bulletin **27** (1984), 185–191.

[Nag35]   T. Nagell, *Solution de quelque problémes dans la théorie arithmétique des cubiques planes du premier genre*, Wid. Akad. Skrifter Oslo I **No.1** (1935).

[Ogg71]   A. Ogg, *Rational points of finite order on elliptic curves*, Inventiones Math. **12** (1971), 105–111.

[Par60]   W. Parry, *On the β-expansions of real numbers*, Acta. Math. Acad. Sci. Hungar. **11** (1960), 401–416.

[Par64]   W. Parry, *Representations for real numbers*, Acta. Math. Acad. Sci. Hungar. **15** (1964), 95–105.

[Pet83]   K. Petersen, *Ergodic Theory*, Cambridge University Press, Cambridge, 1983.

[Phi86]   P. Philippon, *Critères pour l'indépendence algébrique*, Publ. I.H.E.S. **64** (1986), 5–52.

[Pie17]   T.A. Pierce, *Numerical factors of the arithmetic forms $\prod_{i=1}^{n}(1 \pm \alpha_i^m)$*, Annals of Math. **18** (1917), 53–64.

[Pin98]   C. Pinner, *Bounding the elliptic Mahler measure*, Math. Proc. Camb. Phil. Soc. **124** (1998), 521–529.

[Poi01]   H. Poincaré, *Sur les propriétés arithmétiques des courbes algébriques*, J. de Liouville **7** (1901), 166–233.

[Poo89]   A.J. van der Poorten, *Some facts that should be better known, especially about rational functions*, Number Theory and Applications (R.A. Mollin, ed.), Kluwer Academic Publishers, New York, 1989, pp. 497–528.

[Rau85]   U. Rausch, *On a theorem of Dobrowolski about the product of conjugate numbers*, Colloq. Math. **50** (1985), 137–142.

[Ray87]   G.A. Ray, *Relations between Mahler measure and values of L-series*, Canadian Journal of Math. **39** (1987), 694–732.

[Rei88]   M. Reid, *Undergraduate Algebraic Geometry*, Cambridge University Press, Cambridge, 1988.

[Rib95a]   P. Ribenboim, *The Fibonacci numbers and the Arctic Ocean*, Proceedings of the Second Gaussian Symp. Conf. A: Mathematics and Theoretical Physics (München 1994) (M. Behara, R. Fritsch, and R.G. Lintz, eds.), W. de Gruyter, Berlin, 1995, pp. 41–83.

[Rib95b]   P. Ribenboim, *The New Book of Prime Number Records*, 3rd ed., Springer, New York, 1995.

[Rit22]   J.F. Ritt, *Periodic functions with a multiplication theorem*, Transactions of the Amer. Math. Soc. **23** (1922), 16–25.

[Rit23]   J.F. Ritt, *Permutable rational functions*, Transactions of the Amer. Math. Soc. **25** (1923), 399–448.

[Rok64]   V.A. Rokhlin, *Metric properties of endomorphisms of compact commutative groups*, Izv. Akad. Nauk. SSSR Ser. Mat. **28** (1964), 867–874 (Russian), English transl. American Math. Society Transl. Ser. 2 **64** (1967), 244-252.

[Ros88]   M.I. Rosen, *A proof of the Lucas–Lehmer test*, American Math. Monthly **95** (1988), 855–856.

[RS95]   D.J. Rudolph and K. Schmidt, *Almost block independence and Bernoullicity of Z^d actions by automorphisms of compact abelian groups*, Inventiones Math. **120** (1995), 455–488.

[RS97]   G. Rhin and C. Smyth, *On the Mahler measure of the composition of two polynomials*, Acta Arith. **79** (1997), 239–247.

[Rud62]   W. Rudin, *Fourier Analysis on Groups*, John Wiley & Sons, New York, 1962.

[Rud73]   W. Rudin, *Functional Analysis*, McGraw–Hill, New York, 1973.

[Rud74]   W. Rudin, *Real and Complex Analysis*, McGraw–Hill, New York, 1974.

[Sal45]   R. Salem, *Power series with integral coefficients*, Duke Math. J. **12** (1945), 153–172.

[Sar82]   P. Sarnak, *Spectral behaviour of quasi-periodic potentials*, Commun. Math. Phys. **84** (1982), 377–401.

[Sch73]   A. Schinzel, *On the product of the conjugates outside the unit circle of an algebraic number*, Acta Arith. **24** (1973), 385–399, Addendum **26** (1975) 329-331.

[Sch80]   W.M. Schmidt, *Diophantine Approximation*, Lecture Notes in Mathematics, vol. 785, Springer, New York, 1980.

[Sch82]   A. Schinzel, *Selected Topics on Polynomials*, University of Michigan Press, Ann Arbor, 1982.

[Sch93]   A. Schmidt, *Ergodic theory of complex continued fractions*, Number Theory with an emphasis on the Markoff spectrum (A.D. Pollington and W. Moran, eds.), Marcel Dekker, 1993, pp. 215–226.

[Sch95]   K. Schmidt, *Dynamical Systems of Algebraic Origin*, Birkhäuser, Basel, 1995.

[Sie66]   C.L. Siegel, *Über einige Anwendungen Diophantischer Approximationen (1929)*, Collected Works (New York), Springer, 1966, pp. 209–266.

[Sil81]   J.H. Silverman, *Lower bound for the canonical height on elliptic curves*, Duke Math. Journal **48** (1981), 633–648.

[Sil86]   J.H. Silverman, *The Arithmetic of Elliptic Curves*, Springer, New York, 1986.

[Sil94]   J.H. Silverman, *Advanced Topics in the Arithmetic of Elliptic Curves*, Springer, New York, 1994.

[Sin59]   Ja.G. Sinai, *On the concept of entropy of a dynamical system*, Dokl. Akad. Nauk. SSSR **125** (1959), 768–771 (Russian), MR 26#1423.

[Sma67]   S. Smale, *Differentiable dynamical systems*, Bulletin of the Amer. Math. Soc. **73** (1967), 747–817.

[Smy71]   C.J. Smyth, *On the product of conjugates outside the unit circle of an algebraic integer*, Bulletin of the London Math. Soc. **3** (1971), 169–175.

[Smy80]   C.J. Smyth, *On the measure of totally real algebraic integers*, J. Australian Math. Soc. **30** (1980), 137–149.

[Smy81a] C.J. Smyth, *A Kronecker-type theorem for complex polynomials in several variables*, Canadian Math. Bulletin **24** (1981), 447–452.

[Smy81b] C.J. Smyth, *On measures of polynomials in several variables*, Bull. Australian Math. Soc. **23** (1981), 49–63, Corrigendum: G. Myerson and C.J. Smyth, **26** (1982), 317-319.

[Smy81c] C.J. Smyth, *On the measure of totally real algebraic integers II*, Math. Comp. **37** (1981), 205–208.

[Sol98]   R. Solomyak, *On coincidence of entropies for two classes of dynamical systems*, Ergodic Theory and Dynamical Systems **18** (1998), 731–738.

[Sou91]   C. Soulé, *Geometrie d'Arakelov et théorie des nombres transcendants*, Astérisque **198-200** (1991), 355–371.

[Ste77]   C.L. Stewart, *On divisors of Fermat, Fibonacci, Lucas and Lehmer numbers*, Proceedings of the London Math. Soc. **35** (1977), 425–447.

[Ste78a] C.L. Stewart, *Algebraic integers whose conjugates lie near the unit circle*, Bull. Soc. Math. France **106** (1978), 169–176.

[Ste78b] C.L. Stewart, *On a theorem of Kronecker and a related question of Lehmer*, Séminaire de Théorie de Nombres Bordeaux 1977/78, Birkhaüser, Basel, 1978.

[TF61]   H.N.V. Temperley and M.E. Fisher, *Dimer problems in statistical mechanics – an exact result*, Philosophical Magazine **6** (1961), 1061–1063.

[Tho90]   D.J. Thouless, *Scaling for the discrete Mathieu equation*, Commun. Math. Phys. **127** (1990), 187–193.

[TL71]   H.N.V. Temperley and E.H. Lieb, *Relations between the 'percolation' and 'colouring' problem and other graph–theoretical problems associated with regular planar lattices: some exact results for the 'percolation' problem*, Proc. Royal Soc. London Ser. A **322** (1971), 251–280.

[Ves91]   A.P. Veselov, *What is an integrable mapping?*, What is integrability? (V.E. Zakharov, ed.), Springer, New York, 1991, pp. 251–272.

[Ves92]   A.P. Veselov, *Growth and integrability in the dynamics of mappings*, Commun. Math. Phys. **145** (1992), 181–193.

[Vil98]   F. Rodriguez Villegas, *Modular Mahler measures I*, Proceedings of "Topics in Number Theory" (S. Ahlgren, G. Andrews, and K. Ono, eds.), Kluwer Academic Publishers, 1998.

[Vou96]   P. Voutier, *An effective lower bound for the height of algebraic numbers*, Acta Arith. **74** (1996), 81–95.

[Wal80]   M. Waldschmidt, *Sur le produit des conjugués extérieurs au cercle unité d'un entier algébrique*, L'Enseignement Mathématique **26** (1980), 201–209.

[Wal82]   P. Walters, *An Introduction to Ergodic Theory*, Springer, New York, 1982.

[War48]   M. Ward, *Memoir on elliptic divisibility sequences*, Amer. Journal of Math. **70** (1948), 31–74.

[War96]   T. Ward, *Rescaling of Markov shifts*, Acta. Math. Univ. Comenianae **LXV** (1996), 149–157.

[War97]   T. Ward, *An uncountable family of group automorphisms, and a typical member*, Bulletin of the London Math. Soc. **29** (1997), 577–584.

[War98a] T. Ward, *Almost all S–integer dynamical systems have many periodic points*, Ergodic Theory and Dynamical Systems **18** (1998), 471–486.

[War98b] T. Ward, *A family of Markov shifts (almost) classified by periodic points*, Journal of Number Theory **71** (1998), 1–11.

[Wei28] A. Weil, *L'arithmétique sur les courbes algébriques*, Acta Math. **52** (1928), 281–315.

[Wei30] A. Weil, *Sur un théorème de Mordell*, Bull. Sci. Math. **54** (1930), 182–191.

[WW63] E.T. Whittaker and G.N. Watson, *A Course of Modern Analysis*, Cambridge University Press, Cambridge, 1963.

[You86] R.M. Young, *On Jensen's formula and $\int_0^{2\pi} \log|1-e^{i\theta}|d\theta$*, American Math. Monthly **93** (1986), 44–45.

[Yuz67] S.A. Yuzvinskii, *Computing the entropy of a group endomorphism*, Sibirsk. Mat. Ž **8** (1967), 230–239 (Russian), English transl. Siberian Math. J. **8** (1968), 172-178.

[Zag93] D. Zagier, *Algebraic numbers close to both 0 and 1*, Math. Comp. **61** (1993), 485–491.

[Zha92] S. Zhang, *Positive line bundles on arithmetic surfaces*, Annals of Math. **136** (1992), 569–587.

Index

Universitext *(continued)*

Rubel/Colliander: Entire and Meromorphic Functions
Sagan: Space-Filling Curves
Samelson: Notes on Lie Algebras
Schiff: Normal Families
Shapiro: Composition Operators and Classical Function Theory
Simonnet: Measures and Probability
Smith: Power Series From a Computational Point of View
Smoryński: Self-Reference and Modal Logic
Stillwell: Geometry of Surfaces
Stroock: An Introduction to the Theory of Large Deviations
Sunder: An Invitation to von Neumann Algebras
Tondeur: Foliations on Riemannian Manifolds
Wong: Weyl Transforms
Zong: Strange Phenomena in Convex and Discrete Geometry